RAND NATIONAL DEFENSE RESEARCH INSTITUTE

ISSUES WITH

Access to Acquisition Data and Information

IN THE DEPARTMENT OF DEFENSE

Doing Data Right in Weapon System Acquisition

Megan McKernan, Nancy Y. Moore,
Kathryn Connor, Mary E. Chenoweth,
Jeffrey A. Drezner, James Dryden,
Clifford A. Grammich, Judith D. Mele,
Walter Nelson, Rebeca Orrie,
Douglas Shontz, Anita Szafran

Prepared for the Office of the Secretary of Defense

For more information on this publication, visit www.rand.org/t/RR1534

Library of Congress Cataloging-in-Publication Data is available for this publication.
ISBN: 978-0-8330-9755-2

Published by the RAND Corporation, Santa Monica, Calif.
© Copyright 2017 RAND Corporation
RAND® is a registered trademark.

Support RAND
Make a tax-deductible charitable contribution at
www.rand.org/giving/contribute

www.rand.org

Preface

Acquisition data are the foundation of decisionmaking, management, and oversight of the weapon-system acquisition portfolio for the Department of Defense. How to effectively and efficiently spend these dollars has been a top priority for the Better Buying Power initiatives led by the Office of the Secretary of Defense (OSD) and the Under Secretary of Defense for Acquisition, Technology, and Logistics.

OSD asked the RAND Corporation to help identify how available data can help assist defense-acquisition decisionmaking. In particular, we documented where data reside, who can access the data, and who owns the information in 21 information systems. This builds on our earlier work (Riposo et al., 2015, and McKernan et al., 2016) by exploring in more detail the data that support decisionmaking.

This report should be of interest to government acquisition professionals, oversight organizations, and the analytic community.

This research was sponsored by the Acquisition Resources and Analysis (ARA) Directorate within the Office of the Under Secretary of Defense for Acquisition, Technology, and Logistics and was conducted within the Acquisition and Technology Policy Center of the RAND National Defense Research Institute, a federally funded research and development center sponsored by the Office of the Secretary of Defense, the Joint Staff, the Unified Combatant Commands, the Navy, the Marine Corps, the defense agencies, and the defense Intelligence Community.

For more information on the RAND Acquisition and Technology Policy Center, see www.rand.org/nsrd/about/atp or contact the director (contact information is provided on the web page).

Contents

Preface iii

Figures and Tables vii

Summary ix

Acknowledgments xiii

CHAPTER ONE

Introduction 1

Objectives and Approach 2

Organization of this Report 3

CHAPTER TWO

Background on Acquisition Data in the Department of Defense 5

CHAPTER THREE

Lessons from the Commercial Sector on Data Management 11

MDM Benefits 12

MDM Pitfalls 12

CHAPTER FOUR

Background and Findings on Deep Dives of Acquisition Information Systems ... 15

Basic Details on the Acquisition Information Systems 18

Types of Questions These Information Systems Answer 18

Owner, Manager, and Host of the Information System 23

Statute or Policies Requiring Each Information System 23

Characterization of the Data in the Information System 23

Security and Access Restrictions Governing the Information System 30

Characterization of the Users 32

Observations 32

CHAPTER FIVE

Strengths and Challenges of Acquisition Data Information Systems 35

Strengths 35

Challenges 36

CHAPTER SIX
Conclusions and Recommendations 39
Deep-Dive Conclusions 40
Recommendations for Improving the Acquisition Data Environment 41

APPENDIXES
A. Deep-Dive Background 45
B. Additional Detail on Master Data Management 49

Abbreviations 55
References 59

Figures and Tables

Figures

2.1. Functional Business Areas 9
4.1. Acquisition Data Resides Inside and Outside DoD 17
A.1. Reviewed Acquisition Data Systems Throughout the Federal Government and DoD ... 46

Tables

2.1. Information Requirements from DoDI 5000.02 6
4.1. Information Systems Explored in This Research 16
4.2. Basic Details on the Information Systems 19
4.3. Acquisition Data from These Systems Can Answer Many Questions Including
     Those Below 22
4.4. Owner, Manager, and Host of the Information Systems 24
4.5. Policies Requiring or Determining Contents of the Information Systems 25
4.6. Characterization of the Data in the Information Systems 28
4.7. Policies Used to Manage Security for DoD Information Systems 31
4.8. Characterization of the Users 33
A.1. Discussion Questions for Deep Dives 47
B.1. MDM Maturity Model Levels 52

Summary

Acquisition data and information are the foundation for decisionmaking, management, and oversight of weapon-system acquisition programs.[1] They are critical to such initiatives as Better Buying Power and its efforts to improve defense acquisition. Previous RAND Corporation research has explored issues with access to acquisition data and information and the origins and implementation of controlled unclassified information labels and security policy (Riposo et al., 2015; McKernan et al., 2016). The work reported here, conducted for the Acquisition Resources and Analysis Directorate in the Office of the Under Secretary of Defense for Acquisition, Technology, and Logistics, builds on the earlier research by reviewing 21 acquisition-data systems, their origins and uses, and how acquisition data might be improved. This report summarizes background on acquisition data, reviews commercial practices in data management, and provides findings and recommendations related to acquisition data.

Acquisition Data in the Department of Defense

Acquisition data in the Department of Defense (DoD) can be both "structured," that is, immediately identified within an electronic structure, such as a relational database, or "unstructured," that is, data that are not in fixed locations but often in free-form text. Such data are collected for a variety of statutory and regulatory requirements and at all levels from program offices in the services to offices within the Office of the Under Secretary of Defense for Acquisition, Technology, and Logistics in both centralized and decentralized locations. DoD also uses data from other federal information systems outside the department. The data themselves differ in time frame (with some dating back more than five decades), while the information systems that contain these data have different hardware, software, and interfaces.

Technological improvements have helped DoD improve data collection efficiency, quality, aggregation, ease of access and use, archiving, and analysis, among other characteristics; however, many information systems are difficult for users to navigate effectively and can take years to fully understand. Most systems are built for reporting, not analysis.

Commercial Sector Practices

DoD is not alone in the data challenges it faces. Many private-sector firms face similar data-quality, availability, and security challenges. One way they have addressed these is through

---

[1] *Data* and *information* are used throughout this report to discuss both structured and unstructured data and information that assist DoD acquisition.

master data management (MDM), an approach that includes "the infrastructure, tools and best practices for governance of official corporate records that may be scattered across diverse databases and other repositories" and that may help "assure that data has been generated, vetted, processed, protected and transmitted according to a consistent set of policies and controls" (Kobielus, 2006, p. 35). MDM components include those for data quality and validation, governance and management, architecture, and security and ownership.

An appropriate MDM approach may provide standardized, quality data with regulated governance. By enabling periodic checks for data accuracy, MDM can reduce redundancy, increase information quality, improve productivity, simplify processes, improve risk management, make reporting consistent, and improve decisionmaking.

Primary alternatives to MDM are data warehouses, data lakes, and a data-management strategy.[2] Data warehouses store vast quantities of structured data using a multidimensional approach but do not allow data quality control or standardization. Data lakes can store enormous amounts of raw data but will not set up relationships between the data, thereby increasing the likelihood of data "silos." A data-management strategy can mitigate information silos and provide for access but does not guarantee the quality of data governance.

Current DoD Acquisition Information Systems

The 21 information systems we reviewed cover a wide variety of functional business areas, including research and development, requirements, budgeting, contracting, program cost, human capital, and acquisition oversight. Acquisition decisionmakers use these data to answer a wide variety of questions related to defense acquisition. Our review identified the owner, manager, and host of each information system; the types of data in it; the statutory or policy mandate for each; and the questions each may answer.

Some information systems contain data that are deemed authoritative, while other information system managers pull authoritative data into their information systems. These systems had from fewer than 100 to nearly 400,000 users. User composition varies as well, from members of the public seeking information to a select audience of DoD program managers.

Strengths and Weaknesses of Acquisition Data Information Systems

While the data sets vary considerably, many share two great strengths: standardization and collection of acquisition-related information in one place where data can be input, accessed, and analyzed by those needing it. DoD is very large, and acquisitions are accomplished by many different organizations. A centralized system with consistent formats helps improve data quality.

---

[2] A *data warehouse* is "[a] database system that is designed to support data archiving and subsequent analysis and reporting. It contains both current and historical data but does not support the transaction processing that is required for a database handling currently ongoing business interactions" (Butterfield and Ngondi, 2016). A *data lake* "is defined as a massive—and relatively cheap—storage repository, such as Hadoop, that can hold all types of data until it is needed for business analytics or data mining. A data lake holds data in its rawest form, unprocessed and ungoverned" (Violino, 2015). A *data-management strategy* "describes what, why and when to manage an organization's data assets" (Shirude, 2015).

Another strength of some of these systems is that a portion of their data are input electronically, with controls to ensure that key data elements are entered, edited, and checked against historical and other data. Several systems have also been established or improved to facilitate analysis of acquisition information. Most systems were launched to respond to a reporting or oversight requirement, but analysis features have been added to help answer some difficult acquisition questions. Finally, there are two versions of some of these systems, one on classified networks and another on unclassified networks. This enables analysts to work in the environment appropriate for the classification of the information without having to transmit unclassified data from the classified environment to the unclassified environment.

At the same time, there are challenges to using these systems and analyzing their data. One of the main challenges is that their data are only as good as what has been input or provided, particularly when there are no means to verify accuracy. Updates present another challenge; some systems have policies for verifying data and requiring timely updates, but some have had difficulties updating acquisition data in a timely manner. Assuring access to those who need to know, while protecting sensitive data, is also a challenge for most of these systems, whose access procedures vary greatly.

Options for Improving Acquisition Data

To improve the quality of acquisition data, we recommend that DoD

1. **Formalize a data governance function.** Data governance plays a key role in the success of acquisition data management and is something that has also been incorporated in MDM in the commercial sector. In particular, data governance can monitor and enforce the use of acquisition tools and determine the process and structure for the controlling, planning, monitoring, and managing of acquisition data.

2. **Seek to improve the quality and analytic value of its data.** Several information managers told us that data verification and validation are top priorities and that both manual and automated checks have been built into these managers' systems. These practices should be continued and expanded to other systems. One way to improve analytic value is to require all new systems to have user and data entry guides and data dictionaries that describe data elements and their sources. Without such information, users may inadvertently misuse the data. DoD may want to also require system owners to develop and update plans and costs for continuous improvement of data quality and analytic value so that decisions can be made on cost-effective improvements.

3. **Improve its analytical capability.** DoD should do so by continuing to collect both structured and unstructured data but should try to come up with better ways of utilizing the unstructured data it collects. Unstructured data require more resources and different capabilities to be useful for analysis. Both types of formats have an important role in the execution, oversight, and analysis of acquisition programs. However, structured data allow the use of topic metatags, can use strategic algorithms to check quality, maximize drop-down menus, and minimize free text in electronic formats. Moving toward structured data would also allow more standardization of formats for acquisition data and promote sharing between systems.

4. **Continue to develop and train workers to use and improve data.** Answering sophisticated acquisition questions requires analysts with detailed knowledge, access, and experience with numerous data sets, as well as knowledge of how information systems and their data have changed over time. DoD needs to ensure that its workforce is educated and trained to fully understand, analyze, and use existing acquisition-data opportunities. It needs to continue focusing on developing an internal capability to use and improve acquisition data to better understand what data are collected and should be collected and how data can inform DoD decisionmaking.

Acknowledgments

We would like to thank the sponsor of this study: Mark Krzysko, Deputy Director, Acquisition Resources and Analysis (ARA), Enterprise Information, within the Office of the Under Secretary of Defense for Acquisition, Technology, and Logistics (OUSD[AT&L]). We would also like to thank our project monitors—Jeff Tucker, acquisition visibility capability manager, OUSD(AT&L)/ARA, and Joseph Alfano, former Enterprise Information Studies Program Manager, OUSD(AT&L)/ARA—for their guidance and support throughout this study. Also in the Office of the Secretary of Defense (OSD), we thank Robert Flowe, OSD Studies and Federally Funded Research and Development Center Management, OUSD(AT&L)/ARA, who provided us with additional background information that informed our analysis. We appreciate the efforts of Paul DiRenzo to facilitate communication with the Office of Enterprise Information in OUSD(AT&L)/ARA; and to Mark Hogenmiller who assisted with gathering source materials and implementing the output of this research within the Data Opportunities Visualization in the Defense Acquisition Visibility Environment. We also thank the Acquisition Visibility team and all who volunteered their valuable time to describe their points of view on these topics.

We want to thank all the information system experts we interviewed throughout DoD for this report. We promised not to reveal their names, but their input was critical to assembling key details on each system.

We are very grateful to the formal reviewers of this document, William Shelton and Akilah Wallace, who helped improve it through their thorough reviews. We also thank Maria Falvo for her assistance during this effort.

Finally, we would like to thank the director of the RAND Acquisition and Technology Policy Center, Cynthia Cook, and the associate director, Christopher Mouton, for their insightful comments on this research.

CHAPTER ONE

Introduction

Acquisition data lay a foundation for the Under Secretary of Defense for Acquisition, Technology, and Logistics's (USD[AT&L]'s) decisionmaking, management, and oversight of the weapon-system acquisition portfolio for the Department of Defense (DoD).[1] Acquisition data help inform, monitor, and achieve several DoD objectives:

- promoting transparency in spending
- understanding the causes of cost growth
- controlling acquisition system costs
- visualizing the distribution of defense spending
- achieving small-business goals
- identifying and preventing fraud, waste, and abuse
- conducting analyses for improved decisionmaking
- compiling and tracking items in various processes
- archiving decisions.

Figuring out ways to spend taxpayer dollars allocated to DoD effectively and efficiently has been a top priority of the Better Buying Power initiatives led by the Office of the Secretary of Defense (OSD) and USD(AT&L). In the Implementation Directive for Better Buying Power 2.0, USD(AT&L) specifically acknowledged the need to streamline decisionmaking by "promptly acquiring relevant data and directing differences of opinion to appropriate decisionmakers. Our managers cannot be effective if process consumes all of their most precious resource—time" (Kendall, 2013, p. 2).

Currently, many weapon-system acquisition data are collected in response to policy directives, congressional reporting needs, and the need to meet USD(AT&L)'s statutory authorities. These information requirements largely reside in DoD Instruction (DoDI) 5000.02 (2015). This data-management strategy fails to address the complete managerial prerogatives of USD(AT&L) and the Better Buying Power initiatives. Additionally, siloed reporting of acquisition data may not fully support the USD(AT&L) decisionmaking processes. Data requirements have generally been developed from a particular functional perspective, resulting in a data "ecosystem" characterized by individual collections of data that are functionally stovepiped and disjointed, each with different rules for collection, retention, and access.

---

[1] The amount of acquisition data collected is vast and includes such information as the cost of weapon systems (both procurement and operations), technical performance, contracts and contractor performance, and program decision memoranda. These data are critical to the Office of the Under Secretary of Defense for Acquisition, Technology, and Logistics's (OUSD[AT&L]'s) management and oversight of the $1.6 trillion portfolio of major weapon programs.

Objectives and Approach

In earlier work (Riposo et al., 2015), we identified the issues associated with managing and sharing Controlled Unclassified Information (CUI) within DoD. In this work, we examine issues with managing and accessing the sources of that data. We were asked to consider the following:

- What data are available to help assist in defense acquisition decisionmaking?
- Where do acquisition data reside?
- Who can access the information?
- Can we get access to these data for acquisition-related purposes?

To answer these questions, we held targeted discussions with acquisition information system managers and supplemented these discussions with reviews of official policy documentation and other open sources on the information systems and their contents. We also reviewed literature on master data management (MDM) to understand best practices in data management, and augmented our findings with RAND Corporation knowledge about using these data systems. Specifically, we reviewed various federal-wide, OSD-wide, and service-level information systems and their data elements to identify where the data that support current information requirements in DoDI 5000.02 reside. We focused first on a broad look at the enterprise acquisition landscape as a whole then particularly on sources of acquisition information that support USD(AT&L) through the Defense Acquisition Executive Summary (DAES) process and Defense Acquisition Board secretariat, Director, Acquisition Resources and Analysis (ARA). Our sponsor, Deputy Director, ARA, Enterprise Information, provided the list of 21 information systems to examine for this analysis:

- federal-wide
  - System for Award Management (SAM)
  - Federal Funding Accountability and Transparency Act (FFATA) Subaward Reporting System (FSRS)
  - Electronic Subcontracting Reporting System (eSRS)
  - USAspending.gov
  - Federal Procurement Data System—Next Generation (FPDS-NG)
- OSD level
  - Procurement Business Intelligence Service (PBIS)
  - Defense Acquisition Management Information Retrieval (DAMIR)
  - Acquisition Information Repository (AIR)
  - Earned Value Management Central Repository (EVM-CR)
  - Knowledge Management/Decision Support (KM/DS)
  - Unified Research and Engineering Database (URED)
  - DoD Congressional Budget Data Site
  - DoD Congressional Budget Query Site
  - Cost Assessment Data Enterprise (CADE)
  - Defense Automated Cost Information System (DACIMS)
  - DoD Resources Data Warehouse (DRDW)
  - Mechanization of Contract Administration Services (MOCAS)

   – Defense Departmental Reporting System (DDRS)
- service level
   – Army ACQBIZ and Army Acquisition Business Enterprise Portal (AABEP)
   – Air Force System Metric and Reporting Tool (SMART)
   – Navy Research, Development & Acquisition Information System (RDAIS).

Through the discussions with the information managers of the 21 information systems, we also identified the major users of DoD acquisition data and who is providing acquisition data to DoD information systems to inform USD(AT&L) decisionmaking on defense acquisition. Finally, we provided recommendations that would improve the quality of acquisition data, ease of access, efficiency of collection and use, and the ability to link data through common data elements.

Organization of This Report

Chapter Two provides background on acquisition data needs and where the information resides. Chapter Three reviews literature on MDM and some of the best practices in data management. Chapter Four presents some summary background information on "deep dives" we conducted for each information system, while Chapter Five presents some strengths and challenges with the current data environment. Chapter Six integrates the conclusions of our literature review and deep dives and provides recommendations. We also include two appendixes. Appendix A includes additional information on the deep dives, and Appendix B includes additional information on MDM.

CHAPTER TWO

Background on Acquisition Data in the Department of Defense

Acquisition data and information take on a wide variety of forms within DoD and include such information as the cost of weapon systems (both procurement and operations), technical performance, contracts and contractor performance, and program decision memoranda.[1] These data can be characterized as both "structured" and "unstructured."[2] They are critical to the management and oversight of major weapon programs.

These data may be collected for statutory, regulation, policy, or other reasons. DoDI 5000.02 provides a detailed list of "statutory and regulatory requirements at each of the milestones and other decision points during the acquisition process" (DoDI 5000.02, 2015, Encl. 1, pp. 47–58). This does not encompass all the requirements, but is a centralized source for many of them. Table 2.1 summarizes these data requirements and the sources of the requirements. As the table suggests, the data requirements cover a large number of topics. Some measure the cost, schedule, and performance of weapon systems, while others examine testing, cybersecurity, requirements, budgeting, alternatives, and technology readiness.

The information needed to fulfill these requirements resides throughout DoD at all levels, from program offices in the services to various offices within OUSD(AT&L). It can be found in decentralized locations (e.g., individual computers) and centralized locations (e.g., information systems). DoD also uses data from various federal information systems. A plethora of acquisition-related data sources is now available. The data elements within these information systems vary. Some data elements are unique, while others may overlap, depending on different definitions.[3] The time frames for the various data elements are nonstationary, meaning, for example, that one information system has data from 1960 to current, while another may have data only from 2010 to current. Acquisition data are stored on differing platforms and hardware; architectures, software, and interfaces; vendors; and databases. The systems' accessibility and security requirements (depending on the data being stored) also vary. Figure 2.1 shows the various business areas into which OUSD(AT&L)/ARA/EI categorizes the data.

---

[1] We use the terms *data* and *information* throughout this report to discuss both the structured and unstructured data and the information DoD acquisition needs.

[2] According to *PC Magazine*'s website, *structured data* are "Data that can be immediately identified within an electronic structure such as a relational database" ("Encyclopedia," undated); *unstructured data* are: "Data that are not in fixed locations. The term generally refers to free-form text such as in word processing documents, PDF files, e-mail messages, blogs, Web pages and social sites" ("Encyclopedia," undated).

[3] According to *PC Magazine*'s website ("Encyclopedia," undated), a *data element* is: "The fundamental data structure in a data processing system. Any unit of data defined for processing is a data element; for example, ACCOUNT NUMBER, NAME, ADDRESS, AND CITY. A data element is defined by size (in characters) and type (alphanumeric, numeric only, true or false, date, etc.). A specific set of values or range of values may also be part of the definition."

Table 2.1
Information Requirements from DoDI 5000.02

Information Requirement	Source
2366a/b Certification Memorandum	10 U.S. Code (USC) 2366a 10 USC 2366b DoDI 5000.02
Acquisition Decision Memorandum	DoDI 5000.02
Acquisition Program Baseline (APB)	10 USC 2435 10 USC 2433a DoD Directive (DoDD) 5000.01
Acquisition Strategy	Public Law (PL) 107-314, Sec. 803 DoDI 5000.02, Enc. 2, Para. 6a
Affordability Analysis	DoDI 5000.02, Enc. 8, Sec. 3
Analysis of Alternatives (AoA)	40 USC 11312 PL 106-398, Sec. 811 10 USC 2366a
AoA Study Guidance and AoA Study Plan	DoDI 5000.02, Para. 5d(1)(b)
Bandwidth Requirements Review	PL 110-417, Sec. 1047 DoDI 5000.02
Capability Development Document	Chairman of the Joint Chiefs of Staff Instruction (CJCSI) 3170.01H Joint Capabilities Integration and Development System (JCIDS) Manual
Capability Production Document	CJCSI 3170.01H JCIDS Manual
Capstone Threat Assessment	Defense Intelligence Agency (DIA) Directive 5000.200 DIA Instruction 5000.002
Clinger-Cohen Act Compliance	40 USC, Subtitle III PL 106-398, Sec. 811
Concept of Operations Operational Mode Summary Mission Profile	JCIDS Manual
Core Logistics Determination/Core Logistics and Sustaining Workloads Estimate	10 USC 2464 10 USC 2366a 10 USC 2366b PL 112-81, Sec. 801 DoDI 5000.02, Enc. 6, Para. 3d(2)
Cost Analysis Requirements Description	DoDI 5000.02, Enc. 10, Sec. 3 DoD 5000.4-M
Cybersecurity Strategy	PL 106-398, Sec. 811 40 USC 11312 DoDI 8500.01E
Development Request for Proposal Release Cost Assessment	DoDI 5000.02, Enc. 10, Para. 2a(5)
DoD Component Cost Estimate	DoDI 5000.02, Para. 5d(3)(b)2b DoDI 5000.02, Enc. 10, Sec. 2
DoD Component Cost Position	DoDI 5000.02, Enc. 10, Para. 2e
DoD Component Live Fire Test and Evaluation Report	DoDI 5000.02

Table 2.1—Continued

Information Requirement	Source
Director, Operational Test and Evaluation Report on Initial Operational Test and Evaluation	10 USC 2399 10 USC 139
Economic analysis	PL 106-398, Sec. 811 DoDI 7041.3
Exit Criteria	DoDI 5000.02
Frequency Allocation Application (DD Form 1494)	PL 102-538, Sec. 104 47 USC 305 47 USC Ch. 8, Sub. I
Full Funding Certification Memorandum	DoDI 5000.02, Enc. 10, Para. 2f
Independent Cost Estimate	10 USC 2434 10 USC 2334
Independent Logistics Assessment	PL 112-81, Sec. 832 DoDI 5000.02, Enc. 6, Sec. 5
Information Support Plan	DoDI 8330.01 DoDI 8320.02 DoDI 8410.03
Information Technology (IT) and National Security System Interoperability Certification	DoDI 8330.01
Initial Capabilities Document	CJCSI 3170.01H JCIDS Manual
Initial Threat Environment Assessment	DIA Directive 5000.200 DIA Instruction 5000.002
Item Unique Identification Implementation Plan	DoDI 8320.04
Life-Cycle Mission Data Plan	DoDD 5250.01
Life-Cycle Sustainment Plan	DoDI 5000.02, Enc. 6, Sec. 3
Live Fire Test and Evaluation Report	10 USC 2366
Low-Rate Initial Production Quantity	10 USC 2400 DoDI 5000.02, Para. 5d(6)(e)
Manpower Estimate	10 USC 2434
Operational Test Agency Report of OT&E Results	DoDI 5000.02
Operational Test Plan	10 USC 2399 DoDI 5000.02, Enc. 5, Para. 3e
Programmatic Environment, Safety, and Occupational Health Evaluation and National Environmental Policy Act/Executive Order 12114 Compliance Schedule	42 USC 4321–4347 Executive Order 12114
Post Implementation Review	40 USC 11313
Preservation and Storage of Unique Tooling Plan	PL 110-417, Sec. 815
Problem Statement	DoDI 5000.02, Enc. 1, Para. 4

Table 2.1—Continued

Information Requirement	Source
Program Certification to the Defense Business Systems Management Committee	10 USC 2222
Program Protection Plan	DoDI 5200.39 DoDI 5200.44 DoDI 5000.02, Enc. 3, Para. 13a
Replaced System Sustainment Plan	10 USC 2437
Request for Proposal	Federal Acquisition Regulation (FAR) Subpart 15.203
Should Cost Target	DoDI 5000.02, Para. 5d(3)(b)1
Spectrum Supportability Risk Assessment	DoDI 4650.01
System Threat Assessment Report	DoDI 5000.02 DIA Directive 5000.200 DIA Instruction 5000.002
Systems Engineering Plan	DoDI 5000.02, Enc. 3, Sec. 2
Technology Readiness Assessment	PL 111-23, Sec. 205
Technology Targeting Risk Assessment	DoDI 5000.02 DIA Directive 5000.200 DIA Instruction 5000.002
Test and Evaluation Master Plan	DoDI 5000.02, Encl. 4 and 5
Waveform Assessment Application	DoDI 4630.09

SOURCE: DoDI 5000.02, 2015, Enc. 1, pp. 47–58.

Many factors affect how acquisition data and information are collected and stored. Multiple, changing conditions affect the management of acquisition data. Information owners and managers may need to consider whether a current architecture can support additional statutory requirements, administrative changes, or security policy changes. Technological advancements may also be implemented to improve the

- collection efficiency
- quality of the data
- aggregation of the data
- ease of access and use of the information system and its data
- analysis of the data
- archiving data for future analysis or education
- knowledge
- decision process
- analysis.

The same factors can also affect the development of various acquisition information systems. Acquisition information systems were created, evolved, or repurposed to meet data needs and for legitimate reasons (statutory needs, business needs, archiving, etc.). Acquisition information systems and the data they contain may be designed to answer today's current questions but may be inflexible for answering tomorrow's. The systems have been developed with varying

Figure 2.1
Functional Business Areas

SOURCE: OUSD(AT&L)/ARA/EI.
RAND *RR1534-2.1*

architectures and interfaces. The systems also require analysts with cross-system analytic skills, given the differences in how information systems function and in the data within the systems. Systems are also difficult for users to navigate effectively; fully understanding and mastering them can take years of consistent access and use. Most systems are built for reporting, not analysis. Compliance and tracking have been the priorities.

We found that each system has barriers to use. Access procedures are complicated and generally consist of many steps that may not ultimately guarantee access. Access procedures and permissions vary between and sometimes within systems. The federal systems have much data available to the public, but the DoD systems are mostly restricted. New users can have great difficulty establishing and maintaining access (how to, where, who, what?). Full access to acquisition information systems enables analysts to maximize use of data; however, the owners and managers of the data have found that balancing security and access needs is difficult.

CHAPTER THREE

Lessons from the Commercial Sector on Data Management

DoD has a vast collection of acquisition data in many systems managed by different organizations. In managing and securing this collection, DoD faces many of the same challenges that large private-sector firms face. Therefore, we searched journal articles and business publications, selecting the most relevant to review for best practices in data management and data management maturity models for any lessons from the commercial sector that might be of value to DoD.

DoD is not alone in having challenges to data management. By one estimate, one-fourth of firms whose revenues place them in the Fortune 1,000 have inaccurate or incomplete critical data (Gartner, Inc., 2007). Similarly, a survey of 452 "Top 500" corporations and middle-market businesses in the United States, United Kingdom, and Australia found that only 34 percent were highly confident in their own data, and only 18 percent were confident in data received from other organizations ("PricewaterhouseCoopers . . . ," 2004; Swartz, 2007, p. 29). Invariably, this poor data quality decreases efficiency.

And despite incentives to improve information, many businesses continue to struggle, as DoD does, with managing large volumes of data. In a 2015 survey of 1,200 "global C-level execs, vice presidents, directors, managers, and administrative staff," 92 percent of respondents found some aspects of data management to be challenging, and most were "reactive" to data management in some way, meaning that data-quality issues could negatively affect businesses before they are found and stopped (McCafferty, 2015, p. 1; Experian Data Quality, 2015, pp. 6–7).

Data security is of particular concern to DoD, and poor data management also endangers data security. A study of 476 businesses and their susceptibility to risk found that 63 percent do not have a fully mature method for controlling and tracking sensitive data, and 19 percent have no method at all (Trustwave, 2014, p. 4; Kerner, 2014). U.S. defense contractors face similar problems: In 2015, these contractors scored a median 650 on a 900-point scale measuring cyber security, while financial institutions scored 710, and retailers scored 670 (Sternstein, 2015).[1]

One way that enterprises, and possibly DoD, can deal with collecting, storing, accessing, and sharing an exponentially rising amount of data is through what is called MDM. MDM "refers to the infrastructure, tools and best practices for governance of official corporate records that may be scattered across diverse databases and other repositories" and may help "assure that

---

[1] BitSight Technologies' assessments of the industries are aggregated for the sectors in their entirety. The company gathers terabytes of data throughout the internet for indicators of poor security, such as infected machines and insecure configurations, to determine security scores. (See BitSight Technologies, undated.)

data has been generated, vetted, processed, protected and transmitted according to a consistent set of policies and controls" (Kobielus, 2006, p. 35). MDM seeks to overcome "information silos" in which individual departments use different systems that often do not communicate with one another, which is an ongoing problem in DoD. Such silos can lead to inaccurate data as the information shared across systems becomes fragmented and distorted. Unstandardized and unverified manual entry across data silos compounds these problems. For example, the same data element can have different values in multiple locations due to manual entry errors, as DoD has found.

Most research we reviewed agrees on these fundamental components for a MDM approach: data quality, validation, and data governance. Nevertheless, data architecture, security, and ownership are also important in a successful MDM strategy. These aspects should not be considered independently of one other but as supporting or even overlapping.

The remainder of this chapter summarizes the MDM model.

MDM Benefits

A properly employed MDM strategy can provide a single access point to standardized, valid data with regulated governance (Loshin, 2009, p. 8; Wise, 2008, p. 8; Deyerle, 2008, p. 4; Oracle, 2013, p. 3; American Institute of CPAs, 2013, p. 2). By enabling periodic checks for data accuracy, MDM can

- reduce redundancy
- increase information quality
- improve productivity
- simplify processes
- improve risk management
- make reporting consistent
- improve decisionmaking (Loshin, 2009).

These benefits can improve procurement and compliance. Better data mean more-informed investments. A survey of 110 United Kingdom private-sector procurement managers found that 95 percent said data quality was vitally important to their procurement objectives, but 50 percent felt that their data were of low quality (Albert, 2011). Data quality and security can also help an organization meet government and stakeholder standards.

MDM Pitfalls

While MDM can provide many benefits to organizations managing enormous amounts of data, an improperly implemented MDM strategy can cause problems beyond those that initially existed.

First, MDM can fail if it lacks the support of organization executives. MDM may require years of adjustment and continuous upkeep. If organization leaders do not see its value, they may withdraw their support. Initiative leaders will need to show how the standardized information and processes will benefit the entire enterprise (Griffin, 2006). Initiative leaders must be politically savvy as they manage changes from traditional data collection and management (Power, 2008, pp. 24–38).

Second, an organization establishing a MDM strategy must comprehend its enormity. Companies that attempt to standardize all master data elements in a single large initiative often encounter multiple problems. To prevent these, organizations should begin with a narrower scope, such as data about customers that is critical to the enterprise. This allows the business to familiarize itself with the process of data cleansing and to address any issues that may arise on a smaller scale. Although it can be tempting to see different elements as interconnected and therefore requiring a "big bang" approach, such an approach can result in "scope creep," where an enterprise believes it cannot define a master data element like "customer" without first defining another master data element, such as "product" (Griffin, 2006).

Third, an enterprise must establish a governance system before implementing an MDM strategy. MDM cannot run without established governance at its core defining the rules under which it will operate. Attempting to define authorities and processes while implementing MDM will lead to confusion and a lack of standardized operations (Cochrane, 2009, p. 50). While data-management discussions tend to focus on technology, only about 20 percent of an MDM project goes toward technology; the remainder falls under processes, i.e., governance (Brandel, 2010, p. 34).

The most important thing to consider with an MDM strategy is that it is a *process*, not a project. This means it requires continuous upkeep and management to ensure continuous quality and accuracy of data. This, in turn, requires instituting continuous checkpoints for incoming and stored data. MDM is thus not an end but a means of achieving standardized, quality, and accessible data.

Background and Findings on Deep Dives of Acquisition Information Systems

As part of this effort to understand acquisition data opportunities,[1] we conducted "deep dives" on a set of information systems. In this chapter, we summarize the information we gathered. We reviewed 21 federal-wide, OSD-level, and service-level information systems and their data elements to identify where are some of the acquisition data or information that supports current requirements in DoDI 5000.02. We reviewed five federal-level information systems, 13 OSD-level information systems, and three service-level systems (one Army, one Air Force, and one Navy). At least one member of our team had previous knowledge of 11 of the 21 systems and limited prior or current knowledge of at least five others. For the final five systems, no one on the team had knowledge from use. We worked with our sponsor on whether to pursue access to the information systems for this effort, ultimately deciding not to do so.

We did not rely exclusively on access to the information systems to conduct the deep dives. We also collected official documentation, as available, and requested additional materials from those managing the information systems. We had some level of open-source materials for all but two of the 21 systems. Finally, we relied heavily on discussions with the information managers, particularly on the information systems for which we had little or no knowledge and for which no open-source materials were available. We were able to conduct discussions for all but one of the 21 systems. Our results depended on the variety of information we were able to collect.

We verified the deep-dive information with information managers in early 2016 to ensure that it was the latest available. Nevertheless, we found that the information in these systems is constantly changing as policy, technology, and other things change. Consequently, it is best to consult the information systems directly for the most up-to-date information.

Table 4.1 lists each information system we explored, its type, whether open-source descriptive information is available on it, and whether we discussed the system with an information manager.

These information systems are owned and managed by various offices within the federal government and DoD. Figure 4.1 illustrates the placement of these systems within DoD. FPDS-NG, SAM, FSRS, eSRS, and USAspending.gov are outside DoD.

As stated previously, we gathered additional information for these deep dives through discussions with information managers. This information covered the following general subjects:

---

[1] By *data opportunities*, we mean identifying data that can potentially be used for analysis of various defense acquisition questions.

Table 4.1
Information Systems Explored in This Research

Information System	Federal-, OSD-, or Service-Level System	Prior RAND Knowledge[a]	Availability of Descriptive Information[b]	Discussion with Information Manager
SAM	Federal	Yes	Yes	Yes
FSRS	Federal	Yes	Yes	Yes
eSRS	Federal	Yes	Yes	Yes
USAspending.gov	Federal	Yes	Yes	Yes
FPDS-NG	Federal	Yes	Yes	Yes
PBIS	OSD	None	Yes	No
DAMIR	OSD	Yes	Yes	Yes
AIR	OSD	Yes	Yes	Yes
EVM-CR	OSD	Yes	Yes	Yes
KM/DS	OSD	Yes (limited)	Yes	Yes
URED	OSD	None	Yes	Yes
DoD Congressional Budget Data Site	OSD	Yes	Yes	Yes
DoD Congressional Budget Query Site	OSD	Yes	Yes	Yes
CADE	OSD	Yes	Yes	Yes
DACIMS	OSD	Yes (limited)	Yes	Yes
DRDW	OSD	Yes (limited)	Yes	Yes
MOCAS	OSD	Yes (limited)	Yes (limited)	Yes
DDRS	OSD	None	None	Yes
ACQBIZ/AABEP	Services	None	Yes	Yes
SMART	Services	Yes (limited)	Yes	Yes
RDAIS	Services	None	None	Yes

SOURCES: Discussions with information managers, official policy documentation, and other open sources.

[a] "Yes (limited)" means that RAND researchers associated with this study have some limited knowledge from using this information system in the past or present. RAND researchers may also only have access to a portion of the total data in the information systems.

[b] "Yes (limited)" means that RAND researchers had access to a limited amount of descriptive information on the information system.

- basic details on the acquisition information system
- types of questions answered with the information system
- owner, manager, and host of the information system and the data in it
- statute or policies that led to the creation of the information system or that provide the reason for collecting the data in the system
- characterization of the data in the information system
- security and access restrictions governing the information system
- characteristics of the users
- strengths and challenges of the information system or the data in it.

Figure 4.1
Acquisition Data Resides Inside and Outside DoD

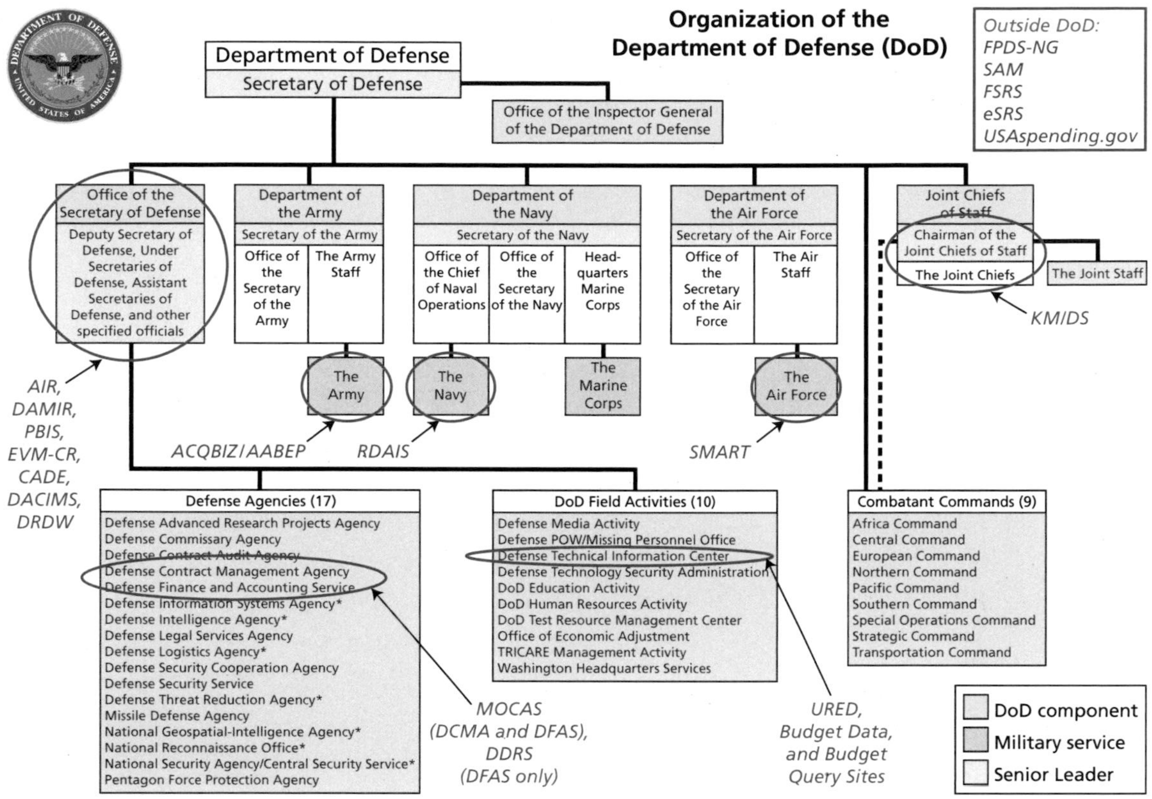

SOURCE: Adapted from DoD organization chart dated March 2012.
*Identified as a combat support agency.
RAND RR1534-4.1

We gathered the following specific information for each of the 21 systems:

- its full name and its abbreviation or common name
- the policies creating or managing the system (e.g., statute, order, policy directive)
- any access restrictions (by data type, user, and security policy)
- the date it entered service
- the openness or availability of the data source
- an overview web address for entering the system
- the "tech stack" or software used for the system
- the functional business area(s) the data support
- the existence of multiple versions because of access issues or attributes
- the purpose of the system and its data
- restrictions on downloading
- the owner, manager, and host of the system
- the process for requesting access
- who owns the data in the system
- data elements (e.g., unit of data with an exact meaning)
- the organization responsible for adding or populating the system

- other information systems that this system feeds
- the organization that developed the system
- the types of questions answered using the data in this system
- authoritative sources for the data in the system
- the users, including their number and composition (e.g., by organization)
- whether the data in the system are considered authoritative
- the strengths of the system and its data
- the data transmitted to or from the system
- the challenges the system and its data face.

Basic Details on the Acquisition Information Systems

For each system, Table 4.2 lists the official abbreviation, the date the system entered service, the URL for the access point for the information system, whether the system is open to the public or is restricted, the functional business area the system supports, and the system's purpose. All the information systems we considered, except MOCAS, have entered service since 1998. Only four of these systems may be accessed by the general public.

These systems cover a wide variety of functional business areas, including

- research and development (R&D)
- requirements
- budgeting
- contracting
- contract performance
- financial execution
- program cost, schedule, and performance
- human capital
- acquisition oversight and portfolio management.

Some systems cover multiple business areas. The Army's ACQBIZ is unique in that it hosts multiple applications, each with its own purpose and business area.

Types of Questions These Information Systems Answer

Decisionmakers and analysts working in defense acquisition need to understand the types of questions that can be answered using the structured and unstructured data in these information systems. They also need to know what questions cannot be answered. We asked information managers to identify some of the questions that can be answered using the data in these information systems. Table 4.3 lists examples of the questions each system can address. The list does not include all questions that may be answered or characterize the degree to which the data can provide satisfactory answers.

Table 4.2
Basic Details on the Information Systems

System	Entered Service	URL	Public or Restricted	Functional Business Area	Purpose(s)
SAM	2000	https://www.sam.gov/portal/SAM/#1	Public	Contracting, grants, small business, finance (i.e., payments), and exclusions activities	Rehosting of all federal Integrated Award Environment systems to preserve previous investments in business logic and business data
FSRS	2010	https://www.fsrs.gov/	Public	Contracting, primarily used for oversight and analysis	Captures and reports prime-contract subawards and executive compensation data to meet FFATA reporting requirements
eSRS	2005	https://www.esrs.gov/	Restricted	Contracting and small business policy	Streamline small-business subcontracting program reporting and provide data for more effective management
USAspending.gov	2007	https://www.usaspending.gov/Pages/Default.aspx	Public	Contracting/financial assistance	Publicly accessible, searchable website to provide public information on how tax dollars are spent
FPDS-NG	2005	https://www.fpds.gov/fpdsng_cms/index.php/en/	Public	Contracting	Provide information on who is procuring, what, when, how, and from whom they are buying, and where work is being done
PBIS	2011	https://reports-osd.altess.army.mil/analytics/saw.dll?bieehome	Restricted	Contracting oversight	Monitor compliance with regulations and policies, support data needs of OSD, Defense Procurement and Acquisition Policy (DPAP), and component leadership, and centralizes the availability and access to procurement data standards, implementation and training tools, and other data
DAMIR	2005	https://ebiz.acq.osd.mil/damir	Restricted	Research, development, test, and evaluation (RDT&E), requirements, budget, contracting, cost spend (finance), schedule/performance, and acquisition oversight	Reporting; storage; quality assurance; analysis; oversight; and tracking cost, schedule, and performance of major acquisition programs

Table 4.2—Continued

System	Entered Service	URL	Public or Restricted	Functional Business Area	Purpose(s)
AIR	2012	https://www.dodtechipedia.mil/AIR	Restricted	Acquisition oversight	Provides one centrally accessible location for all major defense acquisition program (MDAP) and major automated information system (MAIS) acquisition documents in support of oversight and decisionmaking
EVM-CR	2007	http://cade.osd.mil/Background	Restricted	Budget, contracting, cost spend (finance), schedule/performance, and acquisition oversight	Provides a single authoritative source for earned value data for DoD
KM/DS	2003	Not available	Restricted	Requirements	Provides a single, authoritative source of requirements and capabilities documents, and the current status of those documents
URED	2010	https://www.dtic.mil/researchproject/login.html	Restricted	RDT&E, requirements, budget (finance), contracting, schedule, and acquisition oversight	Captures in-progress reporting on DoD-funded research projects performed in DoD labs, academia, or the private sector
Budget Data Site	2007	http://www.dtic.mil/congressional_budget/	Restricted	Budget (finance)	Provides DoD congressional budget data in both PDF and Excel spreadsheet formats
Budget Query Site	2005	https://www.dodtechipedia.mil/dodwiki/display/techipedia/Budget+and+Planning+Information	Restricted	Budget (finance)	Tool for display and query of the President's Budget Request (PBR) data and the congressional-marks data
CADE	2014	http://cade.osd.mil	Restricted	Cost and schedule performance	Analytical tool for integrating cost, schedule, and technical data sources
DACIMS	1999	http://dcarc.cape.osd.mil/CSDR/Dacims.aspx	Restricted	Cost and schedule performance	Provides contract-cost information to support cost estimating
DRDW	1999	Not available	Restricted	Budget (finance)	Provides information regarding program, budget, and acquisition data
MOCAS	1958	https://www.sdw.dcma.mil/	Restricted	Contracting	Supports the administration, management, entitlement of complex contracts, and disbursement of dollars to vendors

Table 4.2—Continued

System	Entered Service	URL	Public or Restricted	Functional Business Area	Purpose(s)
DDRS	1998–1999	https://ddrs.csd.disa.mil/united/html/dr_wp_fp_frontpage.html	Restricted	Budget (finance)	Standardize DoD financial reporting processes
ACQBIZ/AABEP	2012	https://acqdomain.army.mil/	Restricted	RDT&E, requirements, budget (finance), contracting, cost/spend (finance), schedule/performance, acquisition oversight, human capital	Centralized access point for Army acquisition stakeholders to find acquisition business capabilities, data and other information
SMART	2002	https://usaf-acq.platform.milcloud.mil/SMART/smart_app/	Restricted	Acquisition oversight	Support reporting to program executive officers (PEOs), to Assistant Secretary of the Air Force (Acquisition) (SAF/AQ) and to OSD, in particular, monthly acquisition reports and DAES
RDAIS	2009	https://rdais.stax.disa.mil/rdais/	Restricted	Requirements, acquisition oversight	Reporting and tracking system for Navy acquisition programs and authoritative source for programmatic information

SOURCES: Discussions with information managers, official policy documentation, other open sources.

Table 4.3
Acquisition Data from These Systems Can Answer Many Questions Including Those Below

System	Questions
SAM	What are awardee names, locations, contact information, and any special socioeconomic status? What is a contractor's dependence on prime federal contracts, federal subawards? What is the socioeconomic status of subawardees?
FSRS	How much of a prime contract goes to subawards? What are prime contract subawards and executive compensation data? What are the prime contract number, Data Universal Numbering System number of the subcontractor, and amount of subaward, as well as executive compensation? What is a subcontractor's socioeconomic status and dependence on federal or DoD contracts, provided the subcontractor has registered in SAM?
eSRS	Have contractors reported that they met their subcontracting plans for supporting small businesses and meeting socioeconomic goals?
USAspending.gov	How much money does the federal government award by fiscal year? Who is doing the awarding and who is receiving the funding? What is being purchased/awarded?
FPDS-NG	What is the effect of federal procurement on the nation's economy? How have acquisition policy changes and management improvements affected acquisition? What is the impact of full and open competition on the acquisition process? How many sole-source contracts are awarded?
PBIS	Are contracting officers complying with procurement regulations and policy?
DAMIR	How has cost, schedule, and performance changed over time for a particular program or across programs? What is the history of a program? Who is in charge of a program? What is the APB? What is the current cost estimate? How has unit cost changed?
AIR	What acquisition program documentation has been approved for a particular program? What are the OUSD(AT&L)-level acquisition decision memorandums?
EVM-CR	What are the individual metrics of earned value for a particular acquisition program?
KM/DS	What are requirements for systems? Are requirements necessary and feasible for a particular system?
URED	How much money is DoD investing in specific science, research, or technology areas? How many projects are working in the same technology area?
Budget Data	What is contained in current and historical House Armed Services Committee, Senate Armed Services Committee, Authorization Conference Reports, House Appropriations Committee, Senate Appropriations Committee, and Appropriations Conference Reports? How does this affect a particular DoD office?
Budget Query	What is contained in current and historical House Armed Services Committee, Senate Armed Services Committee, Authorization Conference Reports, House Appropriations Committee, Senate Appropriations Committee, and Appropriations Conference Reports? How does this affect a particular DoD office?
CADE and DACIMS	What is the breakout of labor versus materials for a weapon system? For a particular portion of a system, like the airframe? What is a typical overhead rate for a ground/air/sea system?
DRDW	What resources are available for program, budget, and acquisition, including dollars appropriated, manpower, and forces?
MOCAS	What are a program's level of funding, amounts disbursed, and amounts remaining? How many resources does the program still have left to spend?
DDRS	What are obligation rates? What are disbursement rates? Where is money being spent on various categories?
ACQBIZ/AABEP	Typically used to answer questions in regard to managing Army acquisition programs, processes, and oversight, but no specific questions were identified

Table 4.3—Continued

System	Questions
SMART	What is the cost, schedule, and performance of Acquisition Category I–III Air Force acquisition programs?
RDAIS	What is the cost, schedule, and performance of Navy acquisition programs?

SOURCES: RAND discussions with information managers, official policy documentation, other open sources.

Owner, Manager, and Host of the Information System

We also collected information on the owners, managers, and hosts of these systems (Table 4.4). The *owner* is the office responsible for oversight of the information system. The *manager* is responsible for day-to-day operations including approving access and troubleshooting technical issues. The owner and manager are sometimes different organizations, but they typically fall within the same, larger organization, such as OUSD(AT&L) and OSD Cost Assessment and Program Evaluation (CAPE). The *host* of the information system often appears to be an office outside the owner or manager and is typically a contractor for the federal systems. The majority of these DoD systems are hosted by the Army's Acquisition, Logistics and Technology Enterprise Systems and Services (ALTESS), the Defense Information Systems Agency (DISA), OSD CAPE, or the Defense Technical Information Center (DTIC).

Statute or Policies Requiring Each Information System

Most of these systems originated in statute requirements, with the FAR also being a common reason for creating a data system. Some systems originated in policies or memoranda from senior DoD leadership. Table 4.5 details the statutes, regulations, policy, and guidance information managers cited for establishing each system.

Characterization of the Data in the Information System

There is no consensus on whether the data in these systems are authoritative. Some systems, such as FPDS-NG, are authoritative, but others pull data from elsewhere, as, for example, DDRS aggregates financial information from many accounting and financial systems throughout the services and defense agencies.

There is also significant variation in the dates of the data in these information systems. One version of FPDS contains data going back as far as 1951 for DoD. For MOCAS and several other systems, there may be some historical data back to the 1960s. Likewise, there is some variation in whether a formal data dictionary exists and, if one does, whether it is available to users. In some cases, information managers use the data dictionary for planning but do not provide it to users. The federal systems have openly available, formal data dictionaries for the structured data in them. In some systems, data elements have been added over time, or their definitions have changed. For example, Standard Industry Codes, classifications used to determine business size, were replaced in 1997 by North American Industry Classification System

Table 4.4
Owner, Manager, and Host of the Information Systems

Name	Owner	Manager	Host
SAM	General Services Administration (GSA)	GSA	IBM
FSRS	GSA	GSA	Symplicity
eSRS	GSA, but looks to Small Business Administration for policy	GSA	Symplicity
USAspending.gov	Department of the Treasury, Bureau of the Fiscal Service, Office of Financial Innovation and Transformation	Department of the Treasury, Bureau of the Fiscal Service, Office of Financial Innovation and Transformation	Treasury WC2 cloud
FPDS-NG	GSA	FPDS Program Management Office within Integrated Award Environment office	GSA (IBM currently operates and maintains FPDS-NG for GSA)
PBIS	Defense Procurement and Acquisition Policy/ Program Development and Implementation (DPAP/PDI)	DPAP/PDI	Army ALTESS
DAMIR	OUSD(AT&L)/ARA	OUSD(AT&L)/ARA/EI	OUSD(AT&L)/eBusiness Center from DoD Washington Headquarters Services, Enterprise Information Technology Services Directorate Joint Service Provider
AIR	OUSD(AT&L)/ARA	OUSD(AT&L)/ARA/EI	OUSD(AT&L)/ASD(R&E)/DTIC
EVM-CR	OUSD(AT&L)/OASD(A)/ Performance Assessments and Root Cause Analyses (PARCA)	OUSD(AT&L)/OASD(A)/ PARCA	OSD CAPE
KM/DS	Joint Staff J-8 and Joint Requirements Oversight Council (JROC) Secretariat	Joint Staff J-8 and JROC Secretariat	Joint Staff J-6
URED	OUSD(AT&L)/ASD(R&E)	OUSD(AT&L)/ASD(R&E)	OUSD(AT&L)/ASD(R&E)/DTIC
Budget Data Site	OUSD(AT&L)/ASD(R&E)	OUSD(AT&L)/ASD(R&E)/ Deputy Director, Congressional Activities	OUSD(AT&L)/ASD(R&E)/DTIC
Budget Query Site	OUSD(AT&L)/ASD(R&E)/ Dep Dir Congressional Activities	OUSD(AT&L)/ASD(R&E)/ Deputy Director, Congressional Activities	OUSD(AT&L)/ASD(R&E)/DTIC
CADE	OSD CAPE	OSD CAPE	OSD CAPE
DACIMS	OSD CAPE	OSD CAPE	OSD CAPE
DRDW	OSD CAPE	OSD CAPE	OSD CAPE
MOCAS	Defense Contract Management Agency (DCMA) and Defense Finance and Accounting Service (DFAS)	DCMA (65%) and DFAS (35%)	DISA

Table 4.4—Continued

Name	Owner	Manager	Host
DDRS	DFAS	DFAS	DISA
ACQBIZ/AABEP	Assistant Secretary of the Army (Acquisition, Logistics and Technology) (ASA[ALT])/ Strategic Initiatives Group	PM AcqBusiness	Army ALTESS
SMART	Assistant Secretary of the Air Force (Acquisition Integration Capability) (SAF/AQXS)	PEO Business and Enterprise Systems Directorate/ Acquisition Systems Support Branch	DISA
RDAIS	Assistant Secretary of the Navy for Research, Development and Acquisition (ASN[RDA])	ASN(RDA)	DISA

SOURCES: RAND discussions with information managers.

Table 4.5
Policies Requiring or Determining Contents of the Information Systems

Name	Policies
SAM	Federal Acquisition Streamlining Act of 1994 1996 Debt Collection Improvement Act Small-business statutes: HUBZone and 8a FAR at 4.11 and in Title II FAR part 9
FSRS	FFATA from September 2006 FAR part 4 and Title 2
eSRS	Section 8(d) Small Business Act—15 USC 637(d) FAR 19.7/DFARS 219.7, Small Business Subcontracting Program FAR 52.219-8, Utilization of Small Business Concerns FAR 52.219-9/DFARS 52.219-7003, Small Business Subcontracting Plan (Deviation) FAR 52.219-16, Liquidated Damages DFARS 252.219-7004, Small Business Subcontracting Plan (Test) (Deviation) President's Management Agenda for Electronic Government
USAspending.gov	FFATA of 2006 Office of Management and Budget Guidance Digital Accountability and Transparency Act (DATA Act) of 2014
FPDS-NG	PL 93-400 FAR Subpart 4.6
PBIS	DoD's "Strategic Plan For Defense Wide Procurement Capabilities (A Functional Strategy)" (DoD, 2016)
DAMIR	10 USC 2220 (October 13, 1994) 10 USC 2430 (April 21, 1987) 10 USC 2432 (September 8, 1982) 10 USC 2433 (September 8, 1982) 10 USC 2435 (October 18, 1986) DoDI 5000.02 (January 7, 2015) DoDM 5200.01, Vol. 4 (February 24, 2012) DoDI 5400.04 (March 17, 2009) USD(AT&L) Memorandum, Acquisition Program Baselines (APBs) for Major Defense Acquisition Programs (MDAPs) (July 17, 2007) Defense Acquisition Guidebook (2008)
AIR	DoDI 5000.02 (January 7, 2015) USD(AT&L) Kendall, Acquisition Information Repository Implementation Guidance (September 25, 2012)

Table 4.5—Continued

Name	Policies
EVM-CR	10 USC 2220 (October 13, 1994) 10 USC 2430 (April 21, 1987) 10 USC 2438 (2009) Integrated Program Management Report Data Item Description, DI-MGMT-81861 (June 20, 2012) Contractor Sustainment Report DI-FNCL-81831 (May 10, 2011) DoDI 5000.02 Table 8 (January 7, 2015) DoD Earned Value Management System Interpretation Guide (February 18, 2015) DoD Earned Value Management Implementation Guide (October 2006) Integrated Program Management Report Implementation Guide (January, 24, 2013) Over Target Baseline and Over Target Schedule Guide (December 5, 2012) USD(AT&L) Earned Value Management (EVM) Systems Performance, Oversight, and Governance Memo (August 10, 2011) USD(AT&L) The Program Manager's Guide to Integrated Baseline Review Process Memo (June 4, 2003) MIL-STD-881C (October 3, 2011) 2007 OSD Memo authorizing the operational status of the EVM-CR Memo from Director, PARCA EVM on changes to the XML documentation contractors upload to EVM-CR (Kranz, 2014) Memo outlining industry standard guidelines for earned value management systems (Kaminski, 1996)
KM/DS	CJCSI 5123.01G CJCSI 3170.01I
URED	DoD Directive 5134.3 (November 15, 2011) Director, Defense Research and Engineering (DDR&E) Memorandum, "E-gov and Research and Engineering Database" (December 8, 2006) DDR&E Memorandum, "E-gov and Research and Engineering Database FY07 Data Call Submission" (May 18, 2007) DDR&E Memorandum, "Re-engineering R&E Reporting," (September 22, 2010) ASD(R&E) Memorandum, "Unified Research and Engineering Database (URED) 2011 Data Call" (November 15, 2011) DoDI 3200.12, DoD Scientific and Technical Information Program DoDI 3200.14, Vol. 1, Principles and Operational Parameters of the DoD Scientific and Technical Information Program: General Processes (March 14, 2014) Defense Federal Acquisition Regulation Supplement (DFARS) 252.235.7011
Budget Data Site	There was no policy, statute, or regulation involved in creating the system. Rather, the OSD/Deputy Director Congressional Activities verbally requested creation of both the query and data systems.
Budget Query Site	There was no policy, statute, or regulation involved in creating the system. Rather, the OSD/Deputy Director Congressional Activities verbally requested creation of both the query and data systems.
CADE	10 USC 2220 (October 13, 1994) 10 USC 2430 (April 21, 1987) Contract Work Breakdown Structure DI-MGMT-81334D (May 18, 2011) "Cost Data Summary Report" (DD Form 1921) DI-FNCL-81565C (May 18, 2011) "Functional Cost-Hour Report" (DD Form 1921-1) DI-FNCL-81566C (May 18, 2011) "Progress Curve Report" (DD Form 1921-2) DI-FNCL-81567C (May 18, 2011) Contractor Business Data Report (DD Form 1921-3) DI-FNCL-81765B (May 18, 2011) Software Resources Data Reporting: Initial Developer Report and Data Dictionary DI-MGMT-81739B (May 25, 2011) Software Resources Data Reporting: Final Developer Report and Data Dictionary DI-MGMT-81740A (May 18, 2011) Contractor Sustainment Report (DD Form 1921-4) DI-FNCL-81831 (May 10, 2011) DFARS (Sections 234.7100, 234.7101, 252.234-7003, 242.5003, and 252.234-7004 (November 2010) DoD 5000.04-M-1 (November 4, 2011) DoDI 5000.02 (January 7, 2015) OSD CAPE's Operating and Support Cost-Estimating Guide (March 2014) MIL-STD-881C (October 3, 2011)

Table 4.5—Continued

Name	Policies
DACIMS	10 USC 2220 (October 13, 1994) 10 USC 2430 (April 21, 1987) Contract Work Breakdown Structure DI-MGMT-81334D (May 18, 2011) "Cost Data Summary Report" (DD Form 1921) DI-FNCL-81565C (May 18, 2011) "Functional Cost-Hour Report" (DD Form 1921-1) DI-FNCL-81566C (May 18, 2011) "Progress Curve Report" (DD Form 1921-2) DI-FNCL-81567C (May 18, 2011) Contractor Business Data Report (DD Form 1921-3) DI-FNCL-81765B (May 18, 2011) Software Resources Data Reporting: Initial Developer Report and Data Dictionary DI-MGMT-81739B (May 25, 2011) Software Resources Data Reporting: Final Developer Report and Data Dictionary DI-MGMT-81740A (May 18, 2011) Contractor Sustainment Report (DD Form 1921-4) DI-FNCL-81831 (May 10, 2011) DFARS (Sections 234.7100, 234.7101, 252.234-7003, 242.5003, and 252.234-7004 (November 2010) DoD 5000.04-M-1 (November 4, 2011) DoDI 5000.02 (January 7, 2015) OSD CAPE's Operating and Support Cost-Estimating Guide (March 2014) MIL-STD-881C (October 3, 2011)
DRDW	DoDD 5105.84 (May 11, 2012) Future Years Defense Program Improvement Project (late 1990s)
MOCAS	We were unable to find the exact policies or statutes that led to the origin of this information system.
DDRS	Chief Financial Officers Act of 1990 (PL 101-576, 1990) DoD 7000.14-R, Financial Management Regulation, Vol. 3, Ch. 4 (2009, p. 4-3)
ACQBIZ/AABEP	Director, Acquisition Business Systems (SAAL-RB), Assistant Secretary of the Army (Acquisition, Logistics and Technology) (ASA(ALT)), Information Technology Transformation Plan, Vers. 3.0, August 2009. Director, Architecture & Infrastructure, Office of the Department of Defense Chief Information Officer, Department of Defense Enterprise Architecture (EA) Modernization Blueprint/Transition Plan, February 25, 2011.10 USC 2222, Defense Business Systems: Architecture, Accountability, and Modernization (undated) DoDI 5000.02 Army Regulation 70-1
SMART	Memorandum from AFMC and SAF/AQ on May 14, 2002 Air Force Instruction 63-101, Acquisition and Sustainment Life Cycle Management, April 8, 2009
RDAIS	We were unable to obtain the policy that established the need or called for this information system to be created. The system exists in part to fulfill some of the information requirements for the Selected Acquisition Reports that go to Congress, and the DAES Process and APBs that inform Navy leadership and the OUSD(AT&L) for oversight purposes.

SOURCES: RAND discussions with information managers, official policy documentation, other open sources.

codes, which are updated every five years for the Economic Census (see U.S. Census, undated). And the definition of weapon-system codes, which also used to apply to sustainment contracts in FPDS, have only applied to major weapon system acquisitions since 2003.

Table 4.6 lists, for each data system we considered, the information for which it is authoritative, the type of data available, the years of their availability, and information about any data dictionary.

Table 4.6
Characterization of the Data in the Information Systems

Name	Authoritative Source	Type of Data	Availability	Data Element Dictionary
SAM	Contractors and grantees (current and prospective), exclusions	Entity contact, industries, socioeconomic status, tax identification number, and exclusions	DoD since about 1995 Federal since about 1997	Available to all users
FSRS	Subawards <$25,000	Contracting—spending with subcontractors	2010–present	Available to all users
eSRS	Subcontracting plan performance	Contractors' subcontracting dollar allocation by business size and socioeconomic status	2005; replaced paper reports	Available to all users
USAspending.gov	No; aggregator of SAM, FPDS-NG, FSRS, SmartPay, Award Submission Portal assistance	Contracting/financial assistance–spending with non-federal enterprises	2008–present, earlier data can be downloaded	Available to all users
FPDS-NG	Contract actions	Spending with prime contractors	DoD since about 1951 Federal since about 1979; threshold for inclusion varies	Available to all users
PBIS	No; aggregator of Electronic Document Access, FPDS-NG, Wide Area Workflow (WAWF), Contractor Performance Assessment Reporting System (CPARS)	Contracts data: aggregator of EDA, FPDS-NG, WAWF, CPARS	New awards plus 2 years of obligated data	Standard reports are available for users to view, but it is unclear whether a formal data dictionary exists
DAMIR	Selected Acquisition Reports, Selected Acquisition Report Baseline, APB, and Assessments	Lengthy list of program-specific information including cost, schedule, performance for MDAPs/MAIS	1997	Available in the Defense Acquisition Visibility Environment
AIR	Signed and approved acquisition documentation	List of required, approved acquisition program documentation for MDAPs, MAIS programs	Since 2012, contingent on office of primary responsibility uploading documents	Available in the Defense Acquisition Visibility Environment
EVM-CR	Earned value data: Contract Data Requirements List, Integrated Program Management Report, Contract Performance Report, Integrated Master Schedule, and Contract Funds Status Report	Earned value	2008	We were not able to identify whether there is a formal data dictionary

Table 4.6—Continued

Name	Authoritative Source	Type of Data	Availability	Data Element Dictionary
KM/DS	Capability and requirements documents	Requirements documents, from draft to final: Initial Capabilities Document; Capability Development Document; Capability Production Document; Joint Doctrine, Organization, Training, Materiel, Leadership and Education, Personnel, Facilities and Policy Change Recommendation; Capabilities Based Assessment; Integrated Priority List; Joint Urgent Operational Need; JROC meeting calendar	FY 2000 (some historic back to 1960s)	We were not able to identify whether there is a formal data dictionary
URED	Descriptive information on in-process R&D, e.g., spending, topics	Planned, current, completed DoD-funded R&D efforts	Since 2012; DTIC migrated historical information from the Research Summaries database since approximately 1998	Available to all users
Budget Data Site	No; converts budget PDF files to Excel for analysis	PBR, the four mark-ups, and the final conference report from Congress	Budget since 2007	Exists, but it is unclear whether all users are able to view
Budget Query Site	No; converts budget PDF files to Excel for analysis	PBR, the four mark-ups, and the final conference report from Congress	Query since 2005	Exists, but it is unclear whether all users are able to view
CADE	No; aggregates data from authoritative systems	CADE has a combination of elements from the EVM-CR system and DACIMS	Since 2008 and potentially further back depending on document type	Exists, but it is unclear whether all users are able to view
DACIMS	Contractor Cost Data Reports (CCDR), Software Resource Data Reports (SRDR), forward pricing rate (FPR), Cost and Software Data Reporting (CSDR), Contract Work Breakdown Structure (CWBS) dictionaries, and CSDR validation memos	Cost and Schedule Data Reports and Software Resource Data Reports	CCDR, SRDR, FPR documents and legacy Contractor Cost Data (CCD) MDAP and MAIS reports back to 1966	Exists, but it is unclear whether all users are able to view
DRDW	Draft and completed budget information (President's budget) and program objectives memorandum	Detailed, current Future Years Defense Program (from the most recent President's budget or program objectives memorandum) and historic budget data on all DoD program elements	Since 2000 for most, some historic data back to 1960s	Available to all users

Table 4.6—Continued

Name	Authoritative Source	Type of Data	Availability	Data Element Dictionary
MOCAS	Contracts; contract line item numbers (CLINs); required and actual delivery schedules; committed, disbursed, and remaining funds	Contracts; CLINs; required and actual delivery schedules; committed, disbursed, and remaining funds	1960s	None
DDRS	Services and defense agencies are the authoritative sources of financial reporting data	Financial and accounting	No information	Exists but is not available to users
ACQBIZ/AABEP	Depends on the application in ACQBIZ	Currently, 10 Army applications exist in this environment (one example is acquisition oversight data)	Depends on which information system is being accessed	Exists but is not available to users
SMART	Pulls from other authoritative sources, but is the authoritative source for MARs	Cost, schedule, performance data for Air Force acquisition programs	2002	Exists but is not available to users
RDAIS	Programmatic information within the Navy: cost, schedule, and performance	Cost, schedule, and performance data for Navy programs	No information	Exists but is not available to users

SOURCES: RAND discussions with information managers, official policy documentation, other open sources.

Security and Access Restrictions Governing the Information System

All DoD information systems operate in accordance with relevant policies in the DoDD 8000 series (management of DoD IT systems) and security policies for handling classified information. Table 4.7 provides some security policies that are used to manage most of the information covered in this study.

In addition to these security policies, information managers identified several means for controlling access. These include identity-verification measures, need-to-know, and the following other requirements for accessing the systems:

- CAC
- security clearance
- access to the Secure Internet Protocol Router Network (SIPRNet) or Non-Secure Internet Protocol Router Network (NIPRNet)
- SIPRNet token
- classified computing facility
- government sponsor verifying need-to-know
- DD Form 2875, System Authorization Access Request

Table 4.7
Policies Used to Manage Security for DoD Information Systems

Name	Subject	Issuer	Date	Notes
DoDM 5200.01, Vol. 4	DoD Information Security Program: CUI	Under Secretary for Intelligence (USD[I])	February 24, 2012	Implements policy, assigns responsibilities, and provides procedures for the designation, marking, protection, and dissemination of CUI and classified information
DoDI 5200.39, incorporating change 1	Critical Program Information (CPI) Protection Within DoD	USD(I)	December 28, 2010	Establishes policy and assigns responsibilities for the identification and protection of CPI
DoDD 5205.02E	DoD Operations Security (OPSEC) Program	USD(I)	June 20, 2012	Updates policy and responsibilities governing the DoD OPSEC program
DoDI 8320.02	Sharing Data, Information, and IT Services in DoD	DoD Chief Information Officer (CIO)	August 5, 2013	Establishes policies, assigns responsibilities, and prescribes procedures for securely sharing electronic data, information, and IT services and securely enabling the discovery of shared data throughout DoD
DoDI 8500.01	Cybersecurity	DoD CIO	March 14, 2014	Establishes a DoD cybersecurity program to protect and defend DoD information and IT
DoDI 8510.01	Risk Management Framework (RMF) for DoD IT	DoD CIO	March 12, 2014	Established the RMF for DoD IT. Establishes associated cybersecurity policy, and assigns responsibilities for executing and maintaining the RMF. The RMF replaces DoD Information Assurance Certification and Accreditation Process and manages the life-cycle cybersecurity risk to DoD IT
DoDI 8582.01	Security of Unclassified DoD Information on Non-DoD Information Systems	DoD CIO	June 6, 2012	Establishes policy for managing the security of unclassified DoD information on non-DoD information systems
Guide	"Application Security and Development Security Technical Implementation Guide," Vers. 3, Rel. 9	Developed by DISA for DoD	October 24, 2014	Provides the guidance needed to promote the development, integration, and updating of secure applications
Guide	"Enclave Test and Development Security Technical Implementation Guide," Vers. 1, Rel. 1	Developed by DISA for DoD	January 9, 2014	Provides guidance on enclave test and development security

Table 4.7—Continued

Name	Subject	Issuer	Date	Notes
Guide	"DoD Cloud Computing Security Requirements Guide," Vers. 1, Rel. 1	Developed by DISA for DoD	January 12, 2015	Provides requirements for cloud computing
Guide	"DoD Guidebook for Common Access Card (CAC)-Eligible Contractors for Unclassified Network Access"	DPAP	November 21, 2014	Pulls together multiple policies governing network access

SOURCES: USD(I), DoD CIO, DISA, and DPAP.

- signed nondisclosure agreements (between contractor [originator] and nongovernment employee performing IT support or analysis for DoD) when nontechnical proprietary or technical proprietary data are in the information system
- a *.mil* cmail address associated with a CAC
- DoD employment
- final approval from the information system owner.

Not all the information systems required the above information. Most federal-level systems, or at least major portions of them, are open to the public.

Characterization of the Users

The number of users for these information systems varied from fewer than 100 to nearly 400,000. Information managers may count their users as "registered," "active," "average users per month," or "number of users in a particular period." Composition of users also varies widely. Some of the information managers provided high-level statistics (e.g., public, government, DoD), while others provide specific organization names for users. Table 4.8 describes the users for each of the systems we examined.

Observations

The level of detail we were able to compile on each information system and its contents varied considerably and depended on

- RAND-user experience with individual systems
- availability and access to official policy documentation and other materials on the information systems
- interviewee interpretation of discussion questions.

Table 4.8
Characterization of the Users

Name	Number of Users	Composition
SAM	386,148 active	Current/prospective vendors/contractors/grantees inputting/updating data; govt. submitting exclusions, writing/managing contracts/grants, accessing For Official Use Only (FOUO) information; public and federally funded research and development centers (FFRDCs)
FSRS	17,170 active	Prime contractors, grant recipients, government personnel, and groups focused on "good" government business
eSRS	49,110 active	Prime contractors, contracting officers, and small business advisors
USAspending.gov	25,000–30,000	Public, some federal agencies, congressional staffers, data packagers, state and local governments, and the press
FPDS-NG	262,679 active	Federal contracting/acquisition and contractor communities, CAPE, Congress, analysts, general public
PBIS	No data	OSD policy analysts, Component Policy Analysts, Component Chiefs/Execs of Procurement/Contracting, combatant commands and associated procurement personnel
DAMIR	7,212 registered	OSD staff; military departments (MILDEPs); Defense agencies/field activities; combatant commands; FFRDCs; academia; other government (e.g., Congress, Office of Management and Budget, U.S. Government Accountability Office)
AIR	986 registered	OSD staff; Joint Staff; MILDEPs; defense agencies and field activities; PEOs; program managers and program management offices; combatant commanders; FFRDCs; academia
EVM-CR	2,000+	OSD staff; MILDEPs; FFRDCs; support contractors; data providers (can only access own data)
KM/DS	2,500	DoD employees, military service members, DoD contractors, and other government-agency personnel
URED	Average of 243 per month	Air Force, Army, Defense Advanced Research Projects Agency, Defense Logistics Agency, Defense Threat Reduction Agency, Joint Chiefs of Staff, Missile Defense Agency, Navy, and OSD
Budget Data Site	4,000	OSD, services, headquarters, agencies, and comptrollers
Budget Query Site	30 query users in May 2015	OSD, services, headquarters, agencies, and comptrollers
CADE	3,000 total (DACIMS and CADE)	OSD, services, government and contractors, FFRDCs, prime contractors
DACIMS	3,000 total (DACIMS and CADE)	OSD, services, government, prime contractors
DRDW	700	DoD employees, military, contractors with a SIPRnet account and token
MOCAS	8,000	DFAS employees, DCMA, services, contractors (for their data only)
DDRS	1,100+	DoD only: military services, defense agencies, stakeholders and independent public accountant auditors
ACQBIZ/AABEP	5,000–10,000 (depends on budget cycle)	Army users are 99 percent of users: ASA(ALT) staff and PEOs are the main users, contractor users

Table 4.7—Continued

Name	Number of Users	Composition
SMART	5,500	Air Force only: service headquarters staff, PEOs, program offices, contractor support personnel
RDAIS	830	Mix of DoD, contractors (support contractors), audit agency personnel

SOURCES: RAND discussions with information managers.

Given the variation in access to data, as well as differences in data systems, it is somewhat difficult to generalize across them.

Information available through interviews varied as well. Interviewees interpreted protocol questions in varying ways. They also interpreted common terms, such as "owner, user" or "data element, data dictionary," in different ways. This suggests that a common taxonomy would be difficult to implement but may be necessary.

Basic system details were fairly easy to identify and verify. We were also able to compile a variety of potential questions for each information system, although our list is not comprehensive. Understanding both the details of systems and the questions they can answer is critical for decisionmakers.

Identifying system owners, managers, and hosts can also sometimes be difficult because of the sometimes subtle distinctions between owners and managers (e.g., AIR, DAMIR). In other cases, it was easy to verify this information (e.g., CADE, DACIMS) because one office performs all three functions. Some owners, managers, and hosts also changed over time, so it was not always clear who held what role.

The policies that led to the origins of these systems were not always apparent because some of the systems are older, have shifted objectives, and/or have had manager turnover.

Feedback on security, access, and users was very difficult to compare across systems. Security and access were intertwined in discussions, even though there are supposed to be clear origins for policies that require both security and access restrictions. Managers supplied readily available information on user bases, but this information had varying degrees of detail. Some information systems collect data on total users; others collect data on active or registered users; and some provide the average number of users in a period. Some system managers provided us specific information about user characteristics, while others gave us information on broad categories of users.

The enormous amount of information contained in these systems allowed us to compile a large amount of information on data opportunities. For decisionmakers and users of these data, this analysis provides a cursory look at what is available. In the next chapter, we discuss the strengths and challenges of these systems, as well as caveats to using them.

CHAPTER FIVE

Strengths and Challenges of Acquisition Data Information Systems

For each of the data systems we reviewed, we sought to identify strengths and challenges from the perspectives of the information managers and users. This chapter summarizes what we found.

Strengths

Two of the greatest strengths of many of the systems are the standardization and collection of selected acquisition-related information in one place where it can be input, accessed, and analyzed by those who need it. DoD is very large, and many different organizations are involved in the acquisition process. Inputting data into many different systems can be confusing and tedious. Having one centralized system with consistent formats helps improve data quality. Tracking down highly distributed data can also be very time consuming and inefficient. Trying to compare data that are not standardized can be fraught with problems related to differing units and definitions. By consolidating and standardizing acquisition data, these systems make it possible to more easily and efficiently input, access, and analyze the data to provide consistent insights across time and acquisitions. Information managers for several information systems (e.g., FPDS-NG, DAMIR, DDRS, MOCAS, DACIMS) included in this study have spent a lot of time collecting and standardizing information, in some cases over decades, into structured data that have been used for information oversight, decisionmaking, and analysis over time. Other systems (e.g., AIR, KM/DS) have pulled together key information requirements in the form of unstructured data in one central location.

Another strength of some acquisition-data systems is that certain data are input electronically with controls (e.g., through validation checks and business rules) to ensure that key data elements are entered, edited, and cross checked against historical and other data, which improves data quality. Some of the systems we reviewed are the authoritative sources for key data categories (e.g., SAM for prime contractor or grantee information; FPDS-NG for contract-action data; DACIMS for cost data; EVM-CR for earned value data; DAMIR for program-level cost, schedule, and performance data; and DDRS for financial data). Drawing data elements from these authoritative sources helps other acquisition systems improve the quality and consistency of their own.

Several systems have been established (e.g., CADE, PBIS) or improved (e.g., RDAIS, DAMIR, DRDW) to facilitate analysis of acquisition information. Most systems were launched to respond to a reporting or oversight requirement, but analysis has been added. These systems are attempting to pull together many variables in one place for analysis to improve DoD decisionmaking and to save funding that is typically spent to compile information.

Another potential strength is the presence of two versions of some of these systems, one on NIPRNet and the other on SIPRNet. This enables analysts to work in the appropriate environment, based on classification, without having to transmit unclassified data from the classified environment to the unclassified environment. This prevents some error in data transmission and protects classified information. However, because of the need to ensure data consistency over both environments, this is an additional cost information system managers need to consider.

Challenges

One of the main challenges information managers face relates to the quality of the data in their information systems. This depends, in turn, on the quality of data provided and the quality of the input, particularly when there are no means of verifying that it is accurate and has no data-entry errors. Information managers depend greatly on the originators of the data or information for quality of the data. Some acquisition information originates in organizations other than those that collect it. For example, AIR documents are created by the services, CAPE, and other offices within AT&L. Similarly, URED consolidates DoD-funded R&D project data, status, and reports in one central database, with its quality depending on the completeness and accuracy of data inputs from users. Additionally, information on specific acquisition programs or specific types of information (e.g., cost) may vary by who is inputting the information and how each system defines cost.

Updates present another data challenge. Some systems have policies for verifying data and require timely updates. For example, DAMIR has a set schedule for submitting information from the services and has not had major issues with the timing of submissions, but AIR has had some difficulties uploading finalized acquisition documentation in a timely manner.

Assuring access to those who need to know, while at the same time protecting sensitive data, is a challenge for most of these systems. Access procedures vary greatly by system, which burdens personnel needing to access multiple systems. Similarly, nongovernment employees who access proprietary information from CADE and EVM-CR need to sign nondisclosure agreements with the contractor (originators) of the data. Support contractors typically cannot get real-time access to data because of the need for additional nondisclosure agreements. In one instance, the Army has tried to simplify access procedures by creating an environment, ACQBIZ, that hosts multiple Army applications. There is one point of entry, AABEP, which minimizes multiple access procedures.

Another challenge is the inconsistency of understanding of terms. The same term can have different meanings in different acquisition systems, making analyses across systems particularly challenging. Such confusion might be minimized by using data dictionaries or a system to provide authoritative definitions.

Similarly, there is an inconsistency in the formats of the data in these systems. Data formats range from highly structured ones that can be easily analyzed using standard analysis tools to unstructured formats, some of which are also poorly indexed. Conversion of data format, e.g., from PDF to Excel, can also pose an obstacle to analysis.

Some systems, such as SMART and RDAIS, use cloud storage, while MOCAS is using an outdated programming language (e.g., Common Business-Oriented Language [COBOL]) and other systems (e.g., FPDG-NG, DAMIR) have had to undergo major modification to deal with outdated hardware and software. Such variance in hardware and software makes it challenging for users to work across systems, some of which are still using batch processing, while others update in real time.

The usability of some systems could be greatly improved by adding data elements, leveraging authoritative systems, having real-time editing and verification, updating to new platforms, and building additional analytical capability so that the DoD workforce can use these data sets effectively. Indeed, many information managers we interviewed are aware of system challenges and have plans for improvements that are contingent on obtaining the funds to execute them.

We frequently heard that the information managers had a list of desired, backlogged improvements and sometimes critical updates that needed to be made but either lacked the funding or the time to implement them. Organizations are not able to update their information systems for many reasons. One example from our discussions with information managers is that some information systems have been expanded and modified numerous times, which has created a large amount of code. Given the large amount of code, multiple information managers expressed concern about these systems' flexibility to adapt to new capabilities or other uses. They are also concerned about the rapidly changing technological and security environments that force them to spend funds and time patching the information systems.

CHAPTER SIX

Conclusions and Recommendations

Acquisition data and information take on a wide variety of forms within DoD and include such information as the cost of weapon systems (both procurement and operations), technical performance, contracts and contractor performance, and program decision memoranda. These data are collected for a variety of reasons including statutory requirements, regulation, policy, and other reasons.

The information resides throughout all levels of DoD and can be found both in informal, decentralized locations and in formal, centralized locations (e.g., information systems). DoD also uses other federal data residing elsewhere.

Data elements within this plethora of sources may vary. Some data elements are unique, while others may overlap, depending on different definitions. The time frames and sources of these data vary as well.

Multiple, changing conditions affect the management of acquisition data. Information owners and managers may need to consider whether a current architecture can support additional statutory requirements, administrative changes, or security policy changes. Technological advancements may also be implemented to improve collection efficiency, quality, aggregation, and ease of access or use.

The same conditions can also affect the development of various acquisition systems. Acquisition information systems were created, evolved, or repurposed in response to data needs and for legitimate reasons (e.g., statutory needs). Yet the systems are often difficult for users to navigate effectively and can require years of consistent access and use to fully understand and master. Most systems are built for reporting, not analysis, and compliance and tracking have been the priorities. Acquisition information systems and the data they contain might answer current questions but may be inflexible for future ones.

There are also barriers to use of each information system and working across systems. Access procedures are complicated, and users must generally complete many steps to gain access to the information system and its contents. Access procedures and permissions also vary between and sometimes within systems. The federal systems make an abundance of data available to the public, but DoD systems are mostly restricted. New users can have great difficulty establishing and maintaining access. Full access to acquisition information systems enables analysts to maximize use of data but is not practical, given the need to balance security and access.

We found that the private sector struggles with a similar set of problems, with many having inaccurate or incomplete critical data or a lack of confidence in their own data. Despite incentives to improve information, many businesses continue to struggle with managing large volumes of data. MDM offers lessons for such businesses—and may offer insights to DoD as well.

Deep-Dive Conclusions

We compiled information on 21 federal and DoD information systems that contain structured and unstructured acquisition data and information. The level of detail we were able to pull together on each information system and its contents varied considerably depending on

- RAND team user experience with individual systems
- availability and access to official policy documentation and other materials on the information systems
- interviewee interpretation of discussion questions.

There was a wide variety of interpretation of each of the questions in the interview protocol and how these questions pertained to the individual information systems an information manager oversaw. The output of these discussions showed that even common terms, such as "owner, user" and "data element, data dictionary," are subject to interpretation, which suggests that a common taxonomy would be difficult to implement but may be necessary. Basic details were fairly easy to identify and verify. We also pulled together a large variety of potential questions that can be answered using the data in each information system, but the list is neither comprehensive nor an assessment of how well the questions could be answered. Nevertheless, both are critical information for decisionmakers.

Some factual information can be difficult to assess, given subtle distinctions, such as those between owner and manager, in some cases (e.g., AIR, DAMIR). In other cases (e.g., CADE, DACIMS), it was easy to verify information on owners, managers, and hosts, because all three functions are performed by the same office. Yet some owners, managers, and hosts changed over time, so it was not always clear who held what role.

The policies that led to the origins of these systems were not always apparent because some of the systems are older or have morphed from one objective to others or because of turnover in system management personnel. Some information systems included documentation of the policies that led to the systems' creation and determined what data they were to contain (e.g., DAMIR). In other cases, we were given this information during our discussions with system managers.

The information we obtained from managers about security and access and the user base was very difficult to compare across systems. Security and access were intertwined in discussions, even though rules for these are supposed to have distinct origins in statutes and policies that require both types of restrictions. Similarly, the information we received on users varied by number, type, and characteristic.

For each data system we reviewed, we also sought to identify strengths and challenges for the information manager and users. We summarized the major cross-cutting strengths and challenges themes associated with the systems reviewed. The following are some of the major strengths:

- The collection and standardization of selected acquisition-related information into one place allows data to be input, accessed, and analyzed by those needing to use it.
- Electronic data entry that employs such controls as validation checks and business rules to ensure that key data elements are entered, edited, and cross checked against historical and other data improves data quality.
- These systems have been established or improved to answer acquisition questions. The systems are attempting to pull together variables in one place for analysis to improve DoD decisionmaking and to reduce the costs associated with analysts trying to cobble together information.
- The presence of two versions of some systems, one on NIPRNet and the other on SIPRNet, enables analysts to work in the environment appropriate for the classification of the information and avoids the need to move information from a classified to an unclassified environment.

Information managers also face challenges in managing acquisition data, including the following:

- Data quality varies depending on the quality of what is input or provided. Often, there are no means of verifying accuracy.
- Originators must input new data when the data have changed.
- Assuring access to those who need to know while protecting sensitive data is a challenge because access procedures vary greatly by system, burdening those needing to access multiple systems.
- Terminology is inconsistent. The same term can have different meanings in different acquisition systems, which makes analyses across systems particularly challenging.
- Data formats are inconsistent.
- Hardware and software vary greatly.
- Systems need more data elements, leveraging of authoritative systems, real-time editing and verification, and updating to new platforms.
- Desired improvements are backlogged; sometimes, critical updates lack resources for implementation.

Recommendations for Improving the Acquisition Data Environment

Our analysis yielded several recommendations for improving the DoD acquisition-data environment.

Formalize a Data Governance and Data Management Function

To answer DoD's acquisition questions, the USD(AT&L) should consider formalizing a data management and governance function (e.g., a data steward) to oversee data opportunities. Any decision on a data steward would need to consider who could have the authority to institutionalize and implement these changes, given the diversity of data ownership in DoD.

Our discussions with information managers and our literature review on MDM found that data governance plays a key role in the success of acquisition data management. In particular, data governance can monitor and enforce the use of acquisition tools. Data governance also determines the process and structure for authority control, planning, monitoring, and enforcement over data assets (American Institute of CPAs, 2013, p. 4). While data quality and validation focuses on managing individual pieces of data, data governance focuses on data definitions, policies, and processes, including those for data quality and validation. Data governance has two primary data-management objectives: (1) planning and (2) supervision and control.

A data steward function would need to further identify where and what data opportunities exist by maintaining a master list of data or information and authoritative sources. As our research suggests, authoritative sources are not always integrated into information systems, and it is not apparent that developers have a good understanding of all the authoritative sources. There appears to be a movement in that direction (e.g., DAMIR and CADE are now including data from authoritative sources), but DoD should continue to include data from authoritative sources.

The data steward and information managers should proactively solicit ways to improve the value of the data from all categories of users (inputters, overseers, and analysts) to improve final data quality, capability, access, usability, and functionality. This function could also improve understanding of related systems and identify potential opportunities for consolidation (e.g., eSRS, FSRS). ARA/EI has been pulling together data opportunities on the Defense Acquisition Visibility Environment. This is a good place to start for understanding what can be used.

Currently, information managers are working in a siloed environment, which produces a variety of access procedures. A master list of "how to access" guidelines might help users navigate this environment and establish common access procedures and user interfaces.

Improve Data Quality and Its Analytic Value

DoD should require all new systems to have user and data entry guides and data dictionaries that describe data elements and their sources (e.g., directly from another system, from enterprise personnel entering the data). This informs data opportunities and may eliminate duplication. Information managers should try to minimize manual entry whenever possible or provide validation checks. Ideally, an explicit list of authoritative sources for data elements should be available; new systems should be required to use authoritative sources, and older systems should migrate toward them. However, there is no broad agreement on the evaluation criteria for designating a data source as authoritative.

Information managers frequently mentioned that data verification and validation is a top priority and that both manual and automated checks have been built into their systems. Information managers should continue to expand this best practice.

Information managers mentioned that one of their challenges is to continue adding capabilities while complying with the latest security requirements. DoD should require system owners to develop and update plans and budgets for continuous improvement of data quality and analytic value and to document unfunded requirements linked to these improvements.

In terms of improving analytic value, it is also important to link related data to create a common picture (e.g., DRDW to DAMIR to KM/DS) and to use an open application program interface to enable ad hoc queries or access to raw data.

Fully Utilize Both the Structured and Unstructured Data DoD Collects

Current practice is to collect DoD acquisition data in both structured and unstructured formats. To improve DoD's analytical capability, DoD should continue this practice but should try to come up with better ways of utilizing the unstructured data it collects. Unstructured data require more resources and different capabilities to be useful for analysis. Both types of formats have an important role in the execution, oversight, and analysis of acquisition programs. However, structured data is easier to use for analysis. More specifically, structured data

- allow the use of topic metatags
- can use strategic algorithms to check quality
- maximize drop-down menus; minimize free text.

Similarly, a large amount of acquisition information is produced in unstructured formats. Since not all data can be converted into structured formats, DoD needs to identify ways to make unstructured data more useful. Structured data are easy to use once meaning and access have been determined.

Continue to Develop and Train the DoD Workforce to Use/Improve Data

RAND has spent decades using acquisition data to answer difficult questions on a variety of defense acquisition topics. Answering sophisticated acquisition questions requires analysts with detailed knowledge, access, and experience with numerous data sets. The analysts also need knowledge of how the information systems and their data have changed over time to do trend and other analyses. When utilizing very large data sets, robust processing and storage capacity and the skills of research programmers are critical.

DoD needs to ensure that its workforce is educated and trained to fully understand, analyze, and use existing acquisition data opportunities. The acquisition community must have the skills and aptitude to do the same with these data to make decisions. Last, but important, DoD needs to continue to focus on developing an internal, organic capability to use and improve acquisition data to better understand what data are being collected, what data should be collected, and how that information can inform DoD decisionmaking.

Deep-Dive Background

As part of this effort to understand acquisition data opportunities, the RAND study team was asked to conduct deep dives on a set of information systems. We reviewed 21 federal-wide, OSD-level, and service-level information systems and their data elements to identify where the acquisition data that support current information requirements in DoDI 5000.02 reside. The sponsor of this study provided the final list of systems both within DoD and the wider federal government:

- federal
  - SAM
  - FSRS
  - eSRS
  - USAspending.gov
  - FPDS-NG
- OSD
  - PBIS
  - DAMIR
  - AIR
  - EVM-CR
  - KM/DS
  - URED
  - DoD Congressional Budget Data Site
  - DoD Congressional Budget Query Site
  - CADE
  - DACIMS
  - DRDW
  - MOCAS
  - DDRS
- services
  - ACQBIZ/AABEP
  - SMART
  - Navy RDAIS.

Figure A.1 illustrates the offices that are responsible for these information systems.

We gathered information for these deep dives through discussions with information managers, official policy documentation, other open-source information, and RAND experience

Figure A.1
Reviewed Acquisition Data Systems Throughout the Federal Government and DoD

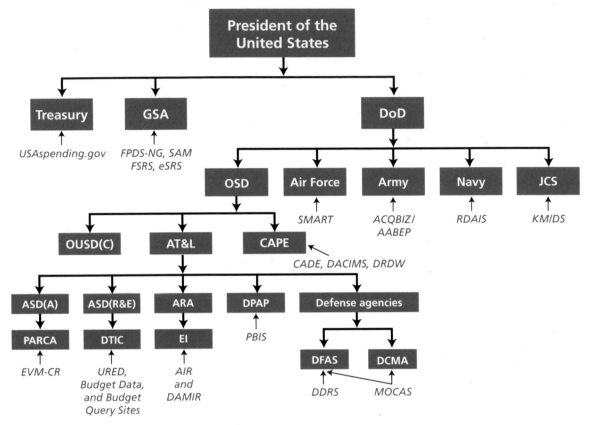

using these information systems. The set of questions in Table A.1 prompted the information we gathered in the discussions.

It is important to note that the information was verified at the beginning of 2016 with the information managers, so the latest available sources were used. However, in our discussions, we found that the information in these systems is constantly changing based on technology changes, policy changes, and other changing needs. It is best to consult the official sources for the most up-to-date information.

Table A.1
Discussion Questions for Deep Dives

Topic	Questions
Background on the Interviewee	What roles have you held in managing and/or using acquisition data?
	How long have you been in your current position?
Basic Details on the Acquisition Information System	What is the acronym or common name for the system?
	What is the full data system name?
	In what year did the information system enter service?
	What URL is the entrance point for the system (NIPR/SIPR)?
	What functional business area(s) does the data source support or fit within (e.g., RDT&E, requirements, budget (finance), contracting, cost and spending (finance), schedule and performance, acquisition oversight, human capital)?
	What is the purpose of the system (e.g., support Defense Acquisition Board process, statutory reason)?
Owner, Manager, and Host of the Information System and Data in that System	What organization owns the information system (e.g., office)?
	What organization manages the information system (e.g., office)?
	What organization hosts the information system?
	What organization owns the data in the information system?
	What organizations are responsible for adding, submitting, and populating the information system?
	What enterprise developed the information system?
	What are the authoritative source(s) for the data in the system?
	Is the data in the system considered to be the authoritative source?
	How is data transmitted to the system?
Security and Access Restrictions Governing the Information System	What policies created the information system or are used to manage it (e.g., statutory, policy directive, etc.)?
	How does an analyst get access?
	What access procedures are in place (e.g., government sponsor needed, CAC-needed, .mil network)?
	What access restrictions are included on the data in the system (e.g., none/publicly releasable, unclassified, FOUO, proprietary)?
	What types of business sensitive information (e.g., personally identifiable information, FOUO, Proprietary, Business Sensitive, Trade Restrictions, Cleared for Public Release) characterize the data source and system protections?
	What access restrictions exist by user (e.g., DoD only, other government, FFRDCs, contractors)?
	Does the security framework distinguish users and user roles between government vs. contractor?
	What access restrictions are in place due to security policy (e.g., CAC needed, .mil network only, SIPR-only)
	Is the data source on the NIPRNet, SIPRNet, or both?
	Characterize the data source's openness or data availability (i.e., does it have outgoing application program interfaces and web services, etc.)?
	What is the "tech stack" (software and software product types) that comprise the data source?
	Are there multiple versions due to access or attributes of the system (e.g., certain data elements are suppressed or delayed in different versions)?
	Are there restrictions on downloading (e.g., time or volume constraint on the amount of data that can be downloaded at one time or during a given period)?
Characterization of the Data in the Information System	What are the data elements in the information system? (Note: this list may be extensive and may not be a simple list of variables.)
	What data systems does this information system pull from? What access restrictions are in place due to security policy (e.g., CAC needed, .mil network only, SIPR-only)?
	What data systems does this information system feed? What access restrictions are in place due to security policy (e.g., CAC needed, .mil network only, SIPR-only)?
	What questions are typically answered using this system (e.g., analytic, oversight, execution)?

Table A.1—Continued

Topic	Questions
Characterization of the Users	How many users use this system?
	What is the composition of users (DoD only; other government, Air Force only, OSD only, FFRDCs, contractors, general public, etc.)?
	Who is the user community by organization (e.g., service headquarters staff, comptroller, Joint Staff, CAPE, international cooperation, manufacturing and industrial base policy, contractors, etc.)?
	Are there additional ways that you are compiling user information than the above?
Strengths and Challenges of the Information System or Data in That System	In your experience, what are the strengths of this system or data in this system (e.g., this is the only place that this type of acquisition data are compiled; the data have been collected and standardized over a very long period of time; unique data elements or group of elements; quality of data)?
	What is the assessed quality, sufficiency, and completeness of the data and information in the source?
	In your experience what are the weaknesses of this system or data in this system (e.g., the data are not thoroughly vetted or is duplicated elsewhere; or the technology governing the system leads to inefficiency or is dated)?
	In your experience, what could be done to improve the value of the information system or quality of data in the system?
	Are you aware of any controls or governance frameworks to improve the quality of the data or usability of the system (e.g., these may be business rules or other)?
Other	Is there anything we have not asked that would help us better understand the information system, quality of data that reside in it, or of access issues associated with it?

Additional Detail on Master Data Management

This appendix provides some additional information on MDM, which was discussed briefly in Chapter Three.

Major MDM Components

Most research we reviewed agrees on two fundamental components for MDM: (1) data quality and validation and (2) data governance. Nevertheless, data architecture, security, and ownership are also important in a successful MDM strategy. These aspects should not be considered independently but as supporting or even overlapping one another.

Data Quality and Validation

Data quality and validation refers to all aspects of ensuring good data in an MDM system. As discussed above, without MDM, data silos can form across departments within the same organization, leading to inaccurate, incomplete, and inconsistent data. Sharpe (2011, p. 32) recommends organizations address data quality before they begin to implement MDM. This can require data stewards who undertake

- data profiling—assessing the quality of data values within a data set and exploring relationships that may exist between collections within and across data sets
- data cleansing—breaking down data components, standardizing the information, and filling in the gaps
- data controls—assessing the potential for the introduction of data flaws
- data monitoring—tracking data quality and correcting variations based on predefined rules (Loshin, 2009; Baum, 2013).

Data-management experts agree that the end goal of data standardization will drive process improvements and future product development, rather than having technology drive the processes (Barlow, 2008, p. 45).

Once an organization cleanses and establishes a mechanism for monitoring data, it must still determine how to receive new quality data. Zoetmulder (2014, p. 10) poses a three-step solution: (1) manage supplier data at the point of entry; (2) manage product data at the point of entry; and (3) if these methods fail, use the analytic tools described above after data entry. Managing supplier or product data at entry will reduce the costs associated with using analytic tools (Zoetmulder, 2014, p. 10).

Data Governance and Management

Data governance determines the process and structure for authority control, planning, monitoring, and enforcement over data assets (American Institute of CPAs, 2013, p. 4). While data quality and validation focuses on managing individual pieces of data, data governance focuses on data definitions, policies, and processes, including those for data quality and validation. Data governance has two primary data-management objectives: (1) planning and (2) supervision and control. These include such activities as reducing operational friction, protecting the needs of data stakeholders, building standard processes, and ensuring transparency (American Institute of CPAs, 2013, p. 7). While governance deals with the corporate rules surrounding data, data management handles the tactical execution of the policies (Beasty, 2008, p. 41).

Before an organization can institute governance, it must understand its information architecture. This includes taking inventory of data assets, assessing how data are managed and used, and evaluating inefficiencies or redundancies. An organization should then document each data function and how it achieves the organization's objectives. Finally, the organization must institute a process framework for information policy (Loshin, 2009, pp. 70–71).

While MDM emphasizes a single point of entry to data and data standardization, data governance does not need to have a single point of control (Reed, 2009). Indeed, organizations should avoid establishing a central data governance operation and instead allow different parts of the organization to manage their own decisions and policies while communicating these to other departments (Reed, 2009, p. 21)—while ultimately determining which method of governance works best for its operations.

Khatri and Brown (2010) outlines five interconnected decision domains related to data assets to capture the elements of successful data governance:

- data principles—direction for all other decisions
- data quality—standards for good data
- metadata—how data will be interpreted
- data access—how data will be accessed
- data life cycle—production, retention, and retirement of data.

This framework can help guide an organization in thinking about what governance covers. Data management will require its own corresponding IT assets to implement the policies. These correspond to differing data assets:

- IT principles—definition of the role of IT within the organization
- IT architecture—organizing logic for data, applications, and infrastructure
- IT infrastructure—foundation for the organization's IT capability
- business application needs—for purchased or internally developed IT applications
- IT investment and prioritization—decisions about where and how much to invest in IT (Weill and Ross, 2004, pp. 27–49).

Ultimately, risk has the greatest effect on governance (Loshin, 2009, p. 72). Lack of oversight over the data and how it is processed can affect the operations and financial accounts of the organization. Poor governance that leads to inadequate data security may also shape compliance issues for protecting sensitive information.

Data Architecture Management

MDM has two architectural components, business and technical (Oracle, 2013), reflecting the distinction between data governance and management. The business component encompasses the applications to manage, clean, and support master data. The technical component profiles, consolidates, and synchronizes the master data across the organization (Oracle, 2013, p. 3).

Organizations have two architectural options (Beasty, 2008). The first is *transaction style*, characterized by a central hub from which the entire organization can access the "golden record" of data definitions and attributes. The second is *federal style*, which follows a distributed model allowing data to reside in appropriate categories with a centralized access point that directs applications (Beasty, 2008, p. 42). Most MDM advocates suggest the transaction style, but the federal style may have benefits for more-fragmented organizations, such as multinational organizations that operate on different databases and systems.

Security and Ownership

Data protection remains a high priority for organizations, especially those that store business confidential information. Poorly written or enforced policies can lead to altered or incomplete data. The organization is responsible for the security of its data and runs risks by not properly securing them. In addition to data usage and ownership, an MDM strategy should address who owns the data, who can access it, and if greater data security is needed.

MDM Alternatives

There are three primary alternative systems to an MDM approach. First, data warehouses are a traditional way to store and manage data.[1] Successful in many enterprises, these systems allow businesses to store vast quantities of structured data using a multidimensional approach (Pedersen, 2009, p. 1). They do not, however, include data quality control, provide standardization, or work with nontraditional, unstructured data (Loshin, 2009, pp. 11–12; Beasty, 2008, p. 40).

Second, data lakes can store enormous amounts of *raw* data for future use (Fitzgerald, 2015).[2] Unlike a data warehouse, a data lake will not model the data, that is, it will not set up relationships between the data, risking the threat of information silos forming across the organization. Like MDM, data lakes require data stewards to govern data quality. This system requires personnel who understand how to manage raw, unstructured data, and to build data relationships as needed (Fitzgerald, 2015, p. 7). Although this mitigates information silos, it creates an additional step that can reduce ease of data access.

Third, an enterprise may undertake what Dearborn (2013) refers to as "a five-step data-management strategy." First, an organization must understand the current data-sharing environment for its business. Second, it must set up a centralized "storage bin" to eliminate data

---

[1] A *data warehouse* is "[a] database system that is designed to support data archiving and subsequent analysis and reporting. It contains both current and historical data but does not support the transaction processing that is required for a database handling currently ongoing business interactions" (Butterfield and Ngondi, 2016).

[2] A *data lake* "is defined as a massive—and relatively cheap—storage repository, such as Hadoop, that can hold all types of data until it is needed for business analytics or data mining. A data lake holds data in its rawest form, unprocessed and ungoverned" (Violino, 2015).

silos and provide quick access to data. Third, it must make information available through a single access point to ease data sharing within the organization. Fourth, it must establish a single gateway for sharing this information with customers. Finally, an organization must be able to share data across networks of other organizations (Dearborn, 2013, p. 15). While this strategy can mitigate information silos and ease access, it does not guarantee data quality or governance.

The MDM Maturity Model

Once an organization establishes a data management system, it may still suffer from disorganized or inaccurate data, particularly if its model lacks sufficient maturity. The maturity model refers to how currently employed management systems can be exploited for an MDM program and for discovering deficiencies that must be remedied to produce a reliable MDM system (Loshin, 2009, p 45). Put another way, the model assesses an organization's current "score" for data management and suggests areas of improvement. The model is flexible; describing what should be done rather than how to do it (CMMI Institute, 2014, p. 4). This allows the model to serve as a ubiquitous benchmark across varying types of organizations.

MDM Maturity Model Assessment Levels

Varying versions of MDM maturity models exist, but most agree on five levels of maturity. Within these five levels, the model tests for contributing factors, including data quality, architecture, usage and ownership, protection, maintenance, and integration. The particular version an organization uses will determine which specific components are tested. Table B.1 lists the five levels of MDM maturity and their characteristics.

Table B.1
MDM Maturity Model Levels

Level	Description
Initial	The lowest level, the organization's data management can be characterized as chaotic or provisional. While it understands issues such as data quality and ownership, the organization suffers from many complications related to them (Aiken et al., 2007, p. 46).
Repeatable, reactive	Although the organization understands it has issues with data management, efforts to remedy this are fairly unsuccessful. Often, new technology may be purchased to solve what is seen as an IT problem instead of comprehending the bigger issue of processes. However, the organization has some ability to replicate successful practices (Loshin, 2009, p. 56).
Defined, managed	Organizations have defined processes that facilitate repeatable processes across divisions. However, these efforts remain in nascent stages (Aiken et al., 2007, p. 46).
Managed, proactive	The organization consistently uses standardized processes through an established governance system across the entire enterprise. These processes are required and monitored to direct data-management efforts (Aiken et al., 2007, p. 46). This allows the business to establish better relationships with customers and suppliers (Loshin, 2009, p. 60).
Optimizing, strategic	The organization operates with reliable data management on the tactical level. It consistently analyzes current processes to determine areas of improvement or necessary change. This reduces operating costs and allows for introduction of services to maintain a competitive edge (Spruit and Pietzka, 2014, p. 5).

Following a case study on developing an MDM maturity model assessment, Spruit and Pietzka (2014, pp. 5–6) identified three broad areas for improving maturity. These are knowledge management, process management, and data landscape management. *Knowledge management* refers to the human capital in an organization. This requires an environment in which knowledge sharing and employee education are encouraged. *Process management* aims to produce lean processes that do not involve unnecessary components but without cutting corners. This speaks to the effectiveness and efficiency of an organization's governance structure: it should run the enterprise without hampering it. Finally, *data landscape management* deals with the construction of the data. This needs to reflect the reality of the business in a reasonable form (Spruit and Pietzka, 2014, pp. 5–6).

Drawbacks to the Maturity Model

While the maturity model presents an informed way for enterprises to manage master data better, its dependence on master data may lead to its irrelevance (Shankar and Menon, 2010). As an organization needs more information produced from outside its enterprise—that is, data it does not own—the need for MDM maturity may slip away. Information collected from publicly available sources reduces the need for data aggregators to inform businesses of their customers' needs and trends. This source of master data is outside the purview of the organization and its MDM. Data managers will need to find a way to integrate such open data within their maturity models for the model to remain relevant (Shankar and Menon, 2010, pp. 24–25).

Abbreviations

AABEP	Army Acquisition Business Enterprise Portal
AIR	Acquisition Information Repository
ALTESS	Acquisition, Logistics and Technology Enterprise Systems and Services
AoA	Analysis of Alternatives
APB	Acquisition Program Baseline
ARA	Acquisition Resources and Analysis Directorate
ASD(A)	Assistant Secretary of Defense for Acquisition
ASD(R&E)	Assistant Secretary of Defense for Research and Engineering
CAC	Common Access Card
CADE	Cost Assessment Data Enterprise
CAPE	Cost Assessment and Program Evaluation
CIO	chief information officer
CJCSI	Chairman of the Joint Chiefs of Staff Instruction
COBOL	Common Business-Oriented Language
CPI	critical program information
CUI	Controlled Unclassified Information
DACIMS	Defense Automated Cost Information System
DAES	Defense Acquisition Executive Summary
DAMIR	Defense Acquisition Management Information Retrieval
DCMA	Defense Contract Management Agency
DDR&E	Director of Defense Research and Engineering
DDRS	Defense Departmental Reporting System
DFARS	Defense Federal Acquisition Regulation Supplement

DFAS	Defense Finance and Accounting Service
DIA	Defense Intelligence Agency
DISA	Defense Information Systems Agency
DoD	Department of Defense
DoDD	Department of Defense Directive
DoDI	Department of Defense Instruction
DPAP	Defense Procurement and Acquisition Policy
DRDW	DoD Resources Data Warehouse
DTIC	Defense Technical Information Center
EI	Enterprise Information office
eSRS	Electronic Subcontracting Reporting System
EVM	Earned Value Management
EVM-CR	Earned Value Management Central Repository
FAR	Federal Acquisition Regulation
FFATA	Federal Funding Accountability and Transparency Act
FFRDC	federally funded research and development center
FPDS-NG	Federal Procurement Data System–Next Generation
FSRS	Federal Funding Accountability and Transparency Act Subaward Reporting System
GSA	General Services Administration
IT	information technology
JCIDS	Joint Capabilities Integration and Development System
JROC	Joint Requirements Oversight Council
KM/DS	Knowledge Management/Decision Support
MAIS	major automated information system
MDAP	major defense acquisition program
MDM	master data management
MILDEP	military department
MIL-STD	military standard
MOCAS	Mechanization of Contract Administration Services

NIPRNet	Non-Secure Internet Protocol Router Network
OPSEC	operations security
OSD	Office of the Secretary of Defense
OUSD(C)	Office of the Under Secretary of Defense (Comptroller)
OUSD(AT&L)	Office of the Under Secretary of Defense for Acquisition, Technology, and Logistics
PARCA	Office of Performance Assessments and Root Cause Analyses
PBIS	Procurement Business Intelligence Service
PBR	President's Budget Request
PDF	Portable Document Format
PDI	Program Development and Implementation
PEO	program executive officer
PL	Public Law
RDAIS	Navy Research, Development & Acquisition Information System
R&D	research and development
RDT&E	Research, development, test, and evaluation
RMF	risk management framework
SAF/AQ	Assistant Secretary of the Air Force (Acquisition)
SAF/AQXS	Assistant Secretary of the Air Force (Acquisition Integration Capability)
SAM	System for Award Management
SIPRNet	Secure Internet Protocol Router Network
SMART	System Metric and Reporting Tool
URED	Unified Research and Engineering Database
URL	Universal Resource Locator (web address)
USC	U.S. Code
USD(AT&L)	Under Secretary of Defense for Acquisition, Technology, and Logistics
WAWF	Wide Area Workflow
WC2	Workplace.gov Community Cloud

References

Aiken, Peter, M. David Allen, Burt Parker, and Angela Mattia, "Measuring Data Management Practice Maturity: A Community's Self-Assessment," *Computer*, Vol. 40, No. 4, April 2007 pp. 42–50. As of February 10, 2017:
https://www.computer.org/csdl/mags/co/2007/04/r4042.html

Air Force Instruction 63-101, *Acquisition and Sustainment Life Cycle Management*, Washington, D.C.: Office of the Assistant Secretary of the Air Force for Acquisition, April 8, 2009.

Albert, Angeline, "Poor Data Quality Hinders Procurement Savings Efforts," *Supply Management*, September 16, 2011. As of July 8, 2015:
http://www.supplymanagement.com/news/2011/poor-quality-data-hinders-procurement-savings-efforts

American Institute of CPAs, *An Overview of Data Management*, 2013. As of October 31, 2016:
https://www.aicpa.org/InterestAreas/InformationTechnology/Resources/DataAnalytics/DownloadableDocuments/Overview_Data_Mgmt.pdf

Barlow, Rick Dana, "Effective Data Management—Which Makes More Sense: Tools to Manage or Managing the Tools?" *Healthcare Purchasing News*, Vol. 32, No. 12, December 2008, pp. 44–47. As of January 23, 2017:
http://www.hpnonline.com/inside/2008-12/0812-PS-DataMgmt.html

Baum, David, Oracle, "Masters of the Data: CIOs Tune in to the Importance of Data Quality, Data Governance, and Master Data Management (MDM)," *CRM Magazine*, Vol. 17, No. 2, February 2013, p. 6. As of January 23, 2017:
http://www.oracle.com/us/c-central/cio-solutions/information-matters/importance-of-data/index.html

Beasty, Colin, "The Master Piece," *CRM Magazine*, Vol. 12, No. 1, January 2008, pp. 39–42.

BitSight Technologies, "The BitSight Security Rating Platform, webpage, undated. As of February 15, 2017:
https://www.bitsighttech.com/security-ratings

Brandel, Mary, "Untangling Your Unruly Data," *Computerworld*, Vol. 44, No. 18, September 27, 2010, pp. 32–35.

Butterfield, Andrew, and Gerard Ekembe Ngondi, eds., *A Dictionary of Computer Science*, 7th ed, online, Oxford University Press, 2016. As of March 28, 2016:
http://www.oxfordreference.com/view/10.1093/acref/9780199688975.001.0001/acref-9780199688975-e-6639

Chairman of the Joint Chiefs of Staff Instruction 3170.01H, "Joint Capabilities Integration and Development System," January 10, 2012.

CMMI Institute, "Why Is Measurement of Data Management Maturity Important?" white paper, September 3, 2014. As of October 31, 2016:
http://whitepapers.dataversity.net/content40478

Cochrane, Mike, "Master Data Management: Avoiding Five Pitfalls of MDM," *Information Management*, Vol. 19, No. 1, January 29, 2009, pp. 49–50.

Dearborn, Justin, "Five Steps to a Complete Enterprise Data Management Strategy," *Health Management Technology*, Vol. 34, No. 11, 2013, pp. 14–15.

Defense Information Systems Agency, "Enclave Test and Development Security Technical Implementation Guide," Vers. 1, Rel. 1, January 9, 2014.

————, "Application Security and Development Security Technical Implementation Guide," Vers. 3, Rel. 9, October 24, 2014.

————, "DoD Cloud Computing Security Requirements Guide," Vers. 1, Rel. 1, January 12, 2015.

Defense Intelligence Agency Directive 5000.200, "Intelligence Threat Support for Major Defense Acquisition Programs," February 1, 2013.

Defense Intelligence Agency Instruction 5000.002, "Intelligence Threat Support for Major Defense Acquisition Programs," February 1, 2013.

Defense Procurement and Acquisition Policy, "DoD Guidebook for Common Access Card (CAC)-Eligible Contractors for Unclassified Network Access," November 21, 2014.

————, "Defense Federal Acquisition Regulation Supplement (DFARS) and Procedures, Guidance, and Information (PGI)," website, February 26, 2016. As of March 21, 2016:
http://www.acq.osd.mil/dpap/dars/dfarspgi/current/

DPAP—*See* Defense Procurement and Acquisition Policy.

Department of Defense 5000.4-M, *Cost Analysis Guidance and Procedures*, December 11, 1992.

———— 7000.14-R, *Financial Management Regulation*, Vol. 3: *General Financial Management Information, Systems, and Requirements*, Ch. 4, "Standard Financial Information Structure," June 2009. As of February 19, 2016:
http://comptroller.defense.gov/Portals/45/documents/fmr/Volume_03.pdf

Department of Defense Directive 5000.01, *The Defense Acquisition System*, May 12, 2003.

———— 5205.02E, *DoD Operations Security (OPSEC) Program*, June 20, 2012.

———— 5250.01, *Management of Intelligence Mission Data (IMD) in DoD Acquisition*, January 22, 2013.

Department of Defense Instruction 3200.12, *DoD Scientific and Technical Information Program (STIP)*, Washington, D.C.: Office of the Secretary of Defense, USD(AT&L), August 22, 2013. As of October 31, 2016:
http://www.dtic.mil/whs/directives/corres/pdf/320012p.pdf

———— 3200.14, *Principles and Operational Parameters of the DoD Scientific and Technical Information Program*, Washington, D.C.: Office of the Secretary of Defense, USD(AT&L), May 13, 1997.

———— 4630.09, *Wireless Communications Waveform Development and Management*, November 3, 2008.

———— 4650.01, *Policy and Procedures for Management and Use of the Electromagnetic Spectrum*, January 9, 2009.

———— 5000.02, *Operation of the Defense Acquisition System*, Washington, D.C.: Office of the Secretary of Defense, USD(AT&L), January 7, 2015.

———— 5200.39, *Critical Program Information (CPI) Protection Within the Department of Defense*, July 16, 2008, incorporating change 1, December 28, 2010.

———— 5200.44, *Protection of Mission Critical Functions to Achieve Trusted Systems and Networks (TSN)*, November 5, 2012.

———— 7041.3, *Economic Analysis for Decisionmaking*, November 7, 1995.

———— 8320.02, *Sharing Data, Information, and Information Technology (IT) Services in the Department of Defense*, August 5, 2013.

———— 8330.01, *Interoperability of Information Technology (IT), Including National Security Systems (NSS)*, May 21, 2014.

———— 8410.03, *Network Management (NM)*, August 29, 2012.

———— 8500.01, *Cybersecurity*, March 14, 2014.

———— 8510.01, *Risk Management Framework (RMF) for DoD IT*, March 12, 2014.

———— 8582.01, *Security of Unclassified DoD Information on Non-DoD Information Systems*, June 6, 2012.

Department of Defense Manual 5200.01, Vol. 4, *DoD Information Security Program: CUI*, February 24, 2012.

Department of Defense Manual 5200.014, Vol. 1, *Principles and Operational Parameters of the DoD Scientific and Technical Information System (STIP): General Processes*, March 14, 2014.

Deyerle, Jeff, "The Benefits of Mature Data Management," *Securities Industry News*, Vol. 20, No. 30, August 25, 2008, pp. 4–7.

Director, Acquisition Business Systems (SAAL-RB), *Assistant Secretary of the Army (Acquisition, Logistics and Technology) (ASA(ALT)), Information Technology Transformation Plan*, Vers. 3.0, August 2009.

Director, Architecture & Infrastructure, Office of the Department of Defense Chief Information Officer, *Department of Defense Enterprise Architecture (EA) Modernization Blueprint/Transition Plan*, February 25, 2011.

DoD—*See* U.S. Department of Defense.

DoDI—*See* Department of Defense Instruction.

"Encyclopedia," *PC Magazine* website, undated. As of November 1, 2016:
http://www.pcmag.com/encyclopedia

Executive Order 12114, "Environmental Effects Abroad of Major Federal Actions," January 4, 1979.

Experian Data Quality, "Create Your Ideal Data Quality Strategy: Become a More Profitable, Informed Company with Better Data Insight," white paper, Boston: Experian Information Solutions, Inc., March 2015. As of July 22, 2015:
https://www.edq.com/globalassets/whitepapers/create-your-ideal-data-quality-strategy.pdf

Federal Acquisition Regulation, Subpart 15.203, Requests for Proposals, 2005.

Fitzgerald, Michael, "Gone Fishing—For Data," interview, *MIT Sloan Management Review*, Vol. 56, No. 3, 2015. As of January 23, 2017:
http://sloanreview.mit.edu/article/gone-fishing-for-data/

Gartner, Inc., "'Dirty Data' is a Business Problem, Not an IT Problem, Says Gartner," press release, Sydney, Australia, March 2, 2007. As of July 22, 2015:
http://www.gartner.com/newsroom/id/501733

Griffin, Jane, "Common Master Data Management Pitfalls to Avoid," *DM Review*, Vol. 16, No. 6, June 2006, p. 8.

Joint Capabilities Integration and Development System (JCIDS) Manual, "Manual for the Operation of the Joint Capabilities Integration and Development System," February 12, 2015.

Kaminski, Paul G., "Industry Standard Guidelines for Earned Value Management Systems," memorandum for Service Acquisition Executives, Assistant Secretary of the Air Force (Financial Management), Director, Ballistic Missile Defense Organization, Director, Defense Contract Audit Agency, Director, Defense Logistics Agency, Director, National Security Agency, December 14, 1996. As of February 12, 2016:
https://dap.dau.mil/_layouts/dap/policysearch.aspx?tag=DoD Policy Letters and Memos&group=Organization s&search=EVM&columns=-Title-DocumentSummary-Keywords

Kendall, Frank, "Implementation Directive for Better Buying Power 2.0—Achieving Greater Efficiency and Productivity in Defense Spending," memorandum for Secretaries of the Military Departments; Deputy Chief Management Officer; Department of Defense Chief Information Officer; Directors of the Defense Agencies; AT&L Direct Reports, Office of the Under Secretary of Defense, April 24, 2013.

Kerner, Sean Michael, "Most Organizations Don't Properly Secure Sensitive Data, Report Finds," *eWeek*, December 12, 2014.

Khatri, Vijay, and Carol V. Brown, "Designing Data Governance," *Communications of the ACM*, Vol. 53, No. 1, January 2010, pp. 148–152.

Kobielus, James, "Master Data Management Is Key to Compliance," *Network World*, Vol. 23, No. 22, June 5, 20065.

Kranz, Gordon, Deputy Director Earned Value Management, "DoD Approved UN/CEFACT XML Documentation Update," Performance Assessments and Root Cause Analyses, memorandum to OSD Earned Value Management Integrated Product Team (EVM IPT), January 31, 2014.

Loshin, David, *Master Data Management*, Science Direct, Elsevier Inc., 2009. As of July 3, 2015:
http://www.sciencedirect.com/science/book/9780123742254

McCafferty, Dennis, "Why Organizations Struggle with Data Quality," *CIO Insight*, March 31, 2015.

McKernan, Megan, Jessie Riposo, Jeffrey A. Drezner, Geoffrey McGovern, Douglas Shontz, and Clifford A. Grammich, *Issues with Access to Acquisition Data and Information in the Department of Defense: A Closer Look at the Origins and Implementation of Controlled Unclassified Information Labels and Security Policy*, Santa Monica, Calif.: RAND Corporation, RR-1476-OSD, 2016. As of December 27, 2016:
http://www.rand.org/pubs/research_reports/RR1476.html

Office of the Under Secretary of Defense for Acquisition, Technology, and Logistics, "EVM Documentation," EVM-CR and Performance Assessments and Root Cause Analyses website, undated. As of February 12, 2016:
http://dcarc.cape.osd.mil/EVM/Documents.aspx

Oracle, "Overview: Oracle Master Data Management," an Oracle White Paper, 2013. As of January 23, 2017:
http://www.oracle.com/us/products/applications/master-data-management/mdm-overview-1954202.pdf

OUSD(AT&L)—*See* Office of the Under Secretary of Defense for Acquisition, Technology, and Logistics.

Pedersen, Torben Bach, "Warehousing the World: A Vision for Data Warehouse Research," in Stanislaw Kozielski and Robert Wrembel, eds., *New Trends in Data Warehousing and Data Analysis*, 2009.

Power, Dan, "The Politics of Master Data Management & Data Governance," *DM Review*, Vol. 18, No. 3, March 2008, pp. 24–38.

"PricewaterhouseCoopers 2004 Data Quality Survey Finds Data Management Strategies Still Not a Top Priority Among Senior Executives, Despite Compliance and Regulatory Drivers That Demand Data Integrity," PR Newswire, November 10, 2004. As of July 22, 2015:
http://www.prnewswire.com/news-releases/pricewaterhousecoopers-2004-data-quality-survey-finds-data-management-strategies-still-not-a-top-priority-among-senior-executives-despite-compliance-and-regulatory-drivers-that-demand-data-integrity-75324372.html

Public Law 101-576, Chief Financial Officers Act of 1990, November 15, 1990. As of February 19, 2016:
http://www.dol.gov/ocfo/media/regs/CFOA.pdf

——— 102-538, The National Telecommunications and Information Organization Act, October 27, 1992.

——— 106-398, Floyd D. Spence National Defense Authorization Act for Fiscal Year 2001, October 30, 2000.

——— 107-314, Bob Stump National Defense Authorization Act for Fiscal Year 2003, December 2, 2002.

——— 110-417, The Duncan Hunter National Defense Authorization Act for Fiscal Year 2009, October 14, 2008.

——— 111-23, Weapon Systems Acquisition Reform Act of 2009, May 22, 2009.

——— 112-81, National Defense Authorization Act for Fiscal Year 2012, December 31, 2011.

Reed, David, "Local Data, Global Master," *Data Strategy*, Vol. 6, No. 3, March 2009, pp. 20–21.

Riposo, Jessie, Megan McKernan, Jeffrey A. Drezner, Geoffrey McGovern, Daniel Tremblay, Jason Kumar, and Jerry M. Sollinger, *Issues with Access to Acquisition Data and Information in the Department of Defense: Policy and Practice*, Santa Monica, Calif.: RAND Corporation, RR-880-OSD, 2015. As of December 24, 2015:
http://www.rand.org/pubs/research_reports/RR880.html

Shankar, Ravi, and Ramesh Menon, "MDM Maturity: Pragmatism, Business Challenges, and the Future of MDM," *Business Intelligence Journal*, Vol. 15, No. 3, 3rd Quarter 2010, pp. 19–25.

Sharpe, Michael, "The Data Mastery Imperative," *InformationWeek*, No. 1305, 2011, pp. 29–32.

Shirude, Sanjay, "Implementing Data Management Strategy Through Data Governance and Data Quality Process," presented to the Portland Metro Chapter of the Data Administration Management Association, July 2015. As of March 28, 2016:
http://www.damapdx.org/index.php?option=com_content&view=article&id=131%3A
jul-2015-chapter-meeting&Itemid=61

Spruit, Marco, and Katharina Pietzka, "MD3M: The Master Data Management Maturity Model," *Computers in Human Behavior*, 2014.

Sternstein, Aliya, "Home Depot Has Better Cyber Security Than 25 US Defense Contractors," *Defense One*, July 6, 2015. As of July 8, 2015:
http://www.defenseone.com/technology/2015/07/home-depot-has-better-cyber-security-25-us-defense-contractors/116995/

Swartz, N., "Data Management Problems Widespread," *Information Management Journal*, Vol. 41, No. 5, 2007, pp. 28–30.

Trustwave, *2014 State of Risk Report*, 2014, pp. 1–20. As of November 1, 2016:
https://www2.trustwave.com/rs/trustwave/images/2014_TW_StateofRiskReport.pdf

U.S. Census, "Economic Census," webpage, undated. As of February 15, 2017:
https://www.census.gov/programs-surveys/economic-census.html

U.S. Code, Title 10, §139, Director of Operational Test and Evaluation, undated. As of November 3, 2016:
https://www.law.cornell.edu/uscode/text/10/139

———, §2222, Defense Business Systems: Architecture, Accountability, and Modernization, undated. As of December 29, 2015:
https://www.law.cornell.edu/uscode/text/10/2222

———, §2334, Independent Cost Estimation and Cost Analysis, undated. As of November 3, 2016:
https://www.law.cornell.edu/uscode/text/10/2334

———, §2366a, Major Defense Acquisition Programs: Determination Required Before Milestone A Approval, undated. As of November 3, 2016:
https://www.law.cornell.edu/uscode/text/10/2366a

———, §2366b, Major Defense Acquisition Programs: Determination Required Before Milestone B Approval, undated. As of November 3, 2016:
https://www.law.cornell.edu/uscode/text/10/2366b

———, §2399, Operational Test and Evaluation of Defense Acquisition Programs, undated. As of November 3, 2016:
https://www.law.cornell.edu/uscode/text/10/2399

———, §2400, Low-Rate Initial Production of New Systems, undated. As of November 3, 2016:
https://www.law.cornell.edu/uscode/text/10/2400

———, §2433a, Critical Cost Growth in Major Defense Acquisition Programs, undated. As of November 3, 2016:
https://www.law.cornell.edu/uscode/text/10/2433a

———, §2434, Independent Cost Estimates, undated. As of November 3, 2016:
https://www.law.cornell.edu/uscode/text/10/2434

———, §2435, Baseline Description, undated. As of November 3, 2016:
https://www.law.cornell.edu/uscode/text/10/2435

———, §2437, Development of Major Defense Acquisition Programs: Sustainment of System to Be Replaced, undated. As of November 3, 2016:
https://www.law.cornell.edu/uscode/text/10/2437

———, §2464, Core Logistics Capabilities, undated. As of November 3, 2016:
https://www.law.cornell.edu/uscode/text/10/2464

U.S. Code, Title 40, Subtitle III, Information Technology Management, undated. As of November 3, 2016:
https://www.law.cornell.edu/uscode/text/40/subtitle-III

———, §11312, Capital Planning and Investment Control, undated. As of November 3, 2016:
https://www.law.cornell.edu/uscode/text/40/11312

———, §11313, Performance and Results-Based Management, undated. As of November 3, 2016:
https://www.law.cornell.edu/uscode/text/40/11313

U.S. Code, Title 42, Ch. 55, National Environmental Policy, §4321, Congressional Declaration of Purpose, undated. As of November 3, 2016:
https://www.law.cornell.edu/uscode/text/42/4321

———, §§4321–4347, National Environmental Policy Act, as amended to October 6, 1992. As of January 23, 2017:
https://www.fsa.usda.gov/Internet/FSA_File/nepa_statute.pdf

U.S. Code, Title 47, §305, Government Owned Stations, undated. As of November 3, 2016:
https://www.law.cornell.edu/uscode/text/47/305

———, Ch. 8, National Telecommunications and Information Administration, Sub. I, Organization and Functions, undated. As of November 3, 2016:
https://www.law.cornell.edu/uscode/text/47/chapter-8/subchapter-I

U.S. Department of Defense, "Strategic Plan for Defense Wide Procurement Capabilities (A Functional Strategy)," Vers. 2.1, February 2016. As of March 4, 2016:
http://www.acq.osd.mil/dpap/pdi/eb/docs/PDI_Functional_Strategy_%285-year_strategic_plan%29_version_2.1.pdf

Violino, Bob, "5 Things CIOs Need to Know About Data Lakes," *CIO*, July 31, 2015. As of March 28, 2016:
http://www.cio.com/article/2948182/big-data/5-things-cios-need-to-know-about-data-lakes.html

Weill, Peter, and Jeanne W. Ross, *IT Governance: How Top Performers Manage IT Decision Rights for Superior Results*, Boston: Harvard Business School Press, 2004.

Wise, Lyndsay, "The Intrinsic Value of Master Data Management," *DM Review*, 2008, pp. 8–10.

Zoetmulder, Eric, "Good Data, Better Procurement," *Summit: The Business of Public Sector Procurement*, Vol. 17, No. 1, Spring 2014, pp. 8–11.